Optimization of Heat and Mass Exchange

Optimization of Heat and Mass Exchange

Special Issue Editors

Brian Agnew
Ivan CK Tam
Xiaojun Shi

MDPI • Basel • Beijing • Wuhan • Barcelona • Belgrade • Manchester • Tokyo • Cluj • Tianjin

Special Issue Editors

Brian Agnew
Newcastle University
UK

Ivan CK Tam
Newcastle Research &
Innovation Institute
Singapore

Xiaojun Shi
Xi'an Jiaotong University
China

Editorial Office
MDPI
St. Alban-Anlage 66
4052 Basel, Switzerland

This is a reprint of articles from the Special Issue published online in the open access journal *Processes* (ISSN 2227-9717) (available at: https://www.mdpi.com/journal/processes/special_issues/heat_mass_exchange).

For citation purposes, cite each article independently as indicated on the article page online and as indicated below:

LastName, A.A.; LastName, B.B.; LastName, C.C. Article Title. *Journal Name* **Year**, *Article Number*, Page Range.

ISBN 978-3-03928-742-0 (Pbk)
ISBN 978-3-03928-743-7 (PDF)

Contents

About the Special Issue Editors

Brian Agnew, Prof. Dr., joined the Department of Mechanical Engineering at Newcastle University in 1984 to focus on heat transfer, internal combustion engines and thermal systems. Subsequently, he was appointed Professor of Energy and the Environment for the School of the Built Environment, Northumbria University. He will continue as a Guest Member of Staff at Newcastle University until September 2022.

Ivan CK Tam, Dr., is an Associate Professor at Newcastle University with a strong track record of leading innovative design projects. His research focus revolves around the combustion process, exhaust emission control, energy management and renewable energy technology. His recent research interest is the application of cryogenic technology on liquefied natural gas and the organic Rankine cycles.

Xiaojun Shi, Dr., is currently an academic in the Department of Mechanical Engineering, Xi'an Jiaotong University. His expertise is in the area of thermal engineering, combined, power, and heat and mass transfer. His recent research is in the area of thermal characteristic testing, modeling of engineering components, and their thermal effects.

 *processes* 

Editorial

Optimization of Heat and Mass Exchange

Brian Agnew [1,*], Ivan CK Tam [2,*] and Xiaojun Shi [3,*]

[1]   NewRail—Newcastle Centre for Railway Research, Newcastle University,
     Newcastle upon Tyne NE1 7RU, UK
[2]   Newcastle Research & Innovation Institute, Newcastle University, 80 Jurong East Street 21,
     #05-04, Singapore 609607, Singapore
[3]   School of Mechanical Engineering, Xi'an Jiaotong University, Xi'an 710049, China
*   Correspondence: brian.agnew@newcastle.ac.uk (B.A.); ivan.tam@newcastle.ac.uk (I.C.T.);
     shixiaojun@xjtu.edu.cn (X.S.)

Received: 4 March 2020; Accepted: 4 March 2020; Published: 9 March 2020

1. Introduction

The needs of society are often a driving force for engineering research. We live in an era in which global warming and atmospheric pollution are of great concern to the extent that they are stimulating research on energy efficiency and means to reduce pollution processes. This is a significant challenge that, if addressed wisely, will have far-reaching effects on sustainable life on our planet. To meet these dual challenges, consideration needs to be given to energy efficiency and pollution reduction through the development of new ideas, processes, and practices that can minimize the limitations imposed by the second law of thermodynamics and of optimisation processes that can extract the largest useful return for energy utilisation.

Heat and mass transfer together with fluid dynamics are three related topics that frequently occur simultaneously in many situations. In fact, the occurrence of only one of these phenomena alone is the exception rather than the rule. The basic equations that describe these phenomena are closely related, and the mathematical techniques for understanding them are very similar. It is therefore sensible to consider these topics together as a unitary subject. This is the approach we have adopted for this Special Issue.

In this Special Issue on "Optimization of Heat and Mass Exchange", we have accepted and published 10 high-quality and original articles. These research papers cover theoretical, numerical, or experimental approaches on heat and mass transport phenomena. The Special Issue operates a rigorous peer-review process with a single-blind assessment and at least two independent reviewers, hence our final acceptance of these published high-quality papers.

2. Papers Presented in the Special Issue

The first paper presented by Sun et al. [1] offers an insight into the compression absorption cascade refrigeration system. The operating conditions are found to have a significant effect on the coefficient of performance, and these parameters are optimized simultaneously in their study. The results show significant quantity of waste heat is recovered to produce a cooling effect required by the system.

Martinez et al. [2] present the application of a mathematical model to temperature and viscosity in a mixture during the crystallization process in a scraped surface heat exchanger. The results are a coupled model based on these parameters of the heat exchanger that allow online estimation with errors below 10% for crystallized system.

Quan et al. [3] performed an analysis on the flow field of port plate pair of an axial piston pump for different piston speeds and inlet fluid velocities with the use of CFD software. It was found that the scale and strength of vortex reduced with an increase in the piston speed. Hence, the energy loss was also reduced and the efficiency was improved.

Su et al. [4] develop a two-dimensional two-channel model to simulate the process of heat transfer during phase change in an unsteady flow passage. The relative parameters of the heat flow inlet section of the corrugated passage are found to reach stability before those of the cold flow inlet section. The simulation reflects the heat transfer mechanism well during phase change in a corrugated flow passage of a plate heat exchanger.

Wang et al. [5] present synergy analysis that reveals the enhancement of heat transfer from the fluid circulation is the most significant at the center of the vortices and at the boundary between them. The work is a numerical investigation of natural convection in a circular enclosure with an internal flat plate with various orientations and aspect ratios. The numerical results show that the width of the plate has almost no effect on the horizontally placed plate.

Faraz et al. [6] apply the similarity transformation technique to a nonlinear partial differential system into a system of order differential equation. The objective is to investigate therm-diffusion effect and multislip effect on an asymmetric casson flow in the presence of chemical reaction. A strong numerical correlation is found from the results with published work in skin-friction-factor and Nusselt number.

Lee et al. [7] investigate forced convection on a plate-fin heat sink in an experimental study. The heat transfer performance and the effect spatial characteristics rate are investigated. They report that the 1-D numerical model with empirical coefficients provide good predicted trends in temperature profiles, thermal resistances and optimal heating length. The characteristics of optimal heating position for both laminar and turbulent flow are investigated in this study.

Alic et al. [8] introduce the computational intelligence methods to estimate heat flux at pool boiling processes in the isolated bubble regime. The performance of computational intelligence methods is determined according to the results of error analysis. The support vector machine regression method is found to perform better than the other methods used for pool boiling heat flux estimation.

Zhou et al. [9] provide an experimental and numerical study in an ice storage tank with finned tubes. The objective is to enhance the performance of the solidification process with the application of axially arranged fins and tubes. The results indicate a remarkable improvement in the performance. Benefits are observed with decreasing initial temperature in refrigerant and water as well.

Lu et al. [10] present a mathematical model and simulation of thermal fields of a large canned induction motor by different calculation methods in the effect of water friction loss. The results are compared with the measurements, obtained from the total losses using the loss separation method. The peak temperature and the temperature distribution of windings are not affected by the water friction loss. This paper will be of special interest to those involved in electrical machines cooling design.

3. Conclusions

We believe that the papers in this Special Issue reveal an exciting area that can be expected to continue to grow in near future, namely, the optimization of mass and heat exchange. The pursuit of work in this area requires expertise in thermal and fluid dynamics, system design, and numerical analysis. We hope that this Issue helps to bring the research community into closer contact with each other. Finally, we would like to thank our authors, reviewers, and editorial staff who have contributed to this Special Issue. I am sure all readers of this Special Issue of *Processes* will find these scientific manuscripts interesting and beneficial to their research work in the coming years.

Author Contributions: All authors contribute equally in this manuscript. All authors have read and agreed to the published version of the manuscript.

Conflicts of Interest: The authors declare no conflict of interest.

References

1.   Quan, X.; Liu, L.; Zhuang, Y.; Zhang, L.; Du, J. Heat Exchanger Network Synthesis Integrated with Compression-Absorption Cascade Refrigeration System. *Processes* **2020**, *8*, 210. [CrossRef]

2. Cruz-Martinez, A.D.; Delgado-Portales, R.E.; Perez-Martinez, J.D.; Gonzalez-Ramirez, J.E.; Villalobos-Lara, A.D.; Borras-Enriquez, A.J.; Moscosa-Santillan, M. Estimation of Ice Cream Mixture Viscosity during Batch Crystallization in a Scraped Surface Heat Exchanger. *Processes* **2020**, *8*, 167. [CrossRef]

3. Quan, L.; Gao, H.; Guo, C.; Che, S. Assessment of Dynamic Flow Field of Port Plate Pair of an Axial Piston Pump. *Processes* **2020**, *8*, 86. [CrossRef]

4. Su, H.; Hu, C.; Gao, Z.; Che, S. Phase Change and Heat Transfer Characteristics of a Corrugated Plate Heat Exchanger. *Processes* **2020**, *8*, 26. [CrossRef]

5. Wang, A.; Ying, C.; Wang, Y.; Yang, L.; Ying, Y.; Zhai, L.; Zhang, W. Effect of Orientation and Aspect Ratio of an Internal Flat Plate on Natural Convection in a Circular Enclosure. *Processes* **2019**, *7*, 905. [CrossRef]

6. Faraz, F.; Imran, S.M.; Ali, B.; Haider, S.; Ying, Y.; Zhai, L.; Zhang, W. Thermo-Diffusion and Multi-Slip Effect on an Axisymetric Casson Flow over a Unsteady Radially Stretching Sheet in the Presence of Chemical Reaction. *Processes* **2019**, *7*, 851. [CrossRef]

7. Lee, J.J.; Kim, H.J.; Kim, D.J.; Haider, S.; Ying, Y.; Zhai, L.; Zhang, W. Thermo-Diffusion and Multi-Slip Effect on an Axisymetric Casson Flow over a Unsteady Radially Stretching Sheet in the Presence of Chemical Reaction. *Processes* **2019**, *7*, 772. [CrossRef]

8. Alic, E.; Das, M.; Kaska, O. Heat Flux Estimation at Pool Boiling Processes with Computational Intelligence Methods. *Processes* **2019**, *7*, 293. [CrossRef]

9. Zhou, H.; Chen, M.; Han, X.; Cao, P.; Yao, F.; Wu, L. Enhancement Study of Ice Storage Performance in Circular Tank with Finned Tubes. *Processes* **2019**, *7*, 266. [CrossRef]

10. Lu, Y.; Mustafa, A.; Rehan, M.A.; Razzaq, S.; Ali, S.; Shahid, M.; Adnan, A.W. The Effect of Water Friction Loss Calculation on the Thermal Field of the Canned Motor. *Processes* **2019**, *7*, 256. [CrossRef]

Article

Heat Exchanger Network Synthesis Integrated with Compression–Absorption Cascade Refrigeration System

Xiaojing Sun, Linlin Liu *, Yu Zhuang, Lei Zhang and Jian Du

Institute of Chemical Process System Engineering, School of Chemical Engineering, Dalian University of Technology, Dalian 116024, Liaoning, China; sunxiaojing0913@mail.dlut.edu.cn (X.S.); zhuangyu@mail.dlut.edu.cn (Y.Z.); keleiz@dlut.edu.cn (L.Z.); dujian@dlut.edu.cn (J.D.)
* Correspondence: liulinlin@dlut.edu.cn

Received: 10 December 2019; Accepted: 6 February 2020; Published: 9 February 2020

Abstract: Compression–absorption cascade refrigeration system (CACRS) is the extension of absorption refrigeration system, which can be utilized to recover excess heat of heat exchanger networks (HENs) and compensate refrigeration demand. In this work, a stage-wise superstructure is presented to integrate the generation and evaporation processes of CACRS within HEN, where the generator is driven by hot process streams, and the evaporation processes provide cooling energy to HEN. Considering that the operating condition of CACRS has significant effect on the coefficient of performance (COP) of CACRS and so do the structure of HEN, CACRS and HEN are considered as a whole system in this study, where the operating condition and performance of CACRS and the structure of HEN are optimized simultaneously. The quantitative relationship between COP and operating variables of CACRS is determined by process simulation and data fitting. To accomplish the optimal design purpose, a mixed integer non-linear programming (MINLP) model is formulated according to the proposed superstructure, with the objective of minimizing total annual cost (TAC). At last, two case studies are presented to demonstrate that desired HEN can be achieved by applying the proposed method, and the results show that the integrated HEN-CACRS system is capable to utilize energy reasonably and reduce the total annualized cost by 38.6% and 37.9% respectively since it could recover waste heat from hot process stream to produce the cooling energy required by the system.

Keywords: HEN synthesis; CACRS; operating condition; MINLP; optimization

---

1. Introduction

Exchanging heat between cold and hot process streams through heat exchangers is an effective form of waste heat recovery. A great many of researches have been undertaken, such as the optimization design of heat exchanger, and the synthesis of heat exchanger networks. For the optimization design of heat exchanger, the nanomaterials were mixed in the pure fluid since the nanoparticles have higher thermal conductivity and can give greater characteristics of carrier fluid [1]. In addition, the turbulent flow of nanofluid in a pipe has been modeled in order to reach better design in view of second law [2]. Also, the researches on heat exchanger network (HEN) synthesis have received considerable development, and still bolstered by new ideas. Mathematical programming approach based on superstructure is one of the most widely used methods for HEN synthesis. In recent years, the original stage-wise superstructure proposed by Yee and Grossmann [3] has been improved and perfected. Such as the assumption of isothermal mixing was shattered [4], and the superstructure was enhanced to contain more possibilities [5], as well as to consider the parallel connection of utilities in stream split branches [6] and series connection of utilities in substages [7].

Absorption refrigeration cycle (ARC) is a high-efficiency energy cycle for heat recovery from middle- and low-temperature heat sources in term of generating cooling energy, and many researches have been conducted [8,9]. For greater energy-saving potential, some studies have also been launched on the integration of ARC with HEN. Ponce-Ortega et al. [10] investigated the combination of ARC and HEN by using Pinch-based and mathematical programming methods, which can recycle heat below the pinch point for refrigerating. Lira et al. [11] presented a superstructure having heat exchange system combined with solar absorption refrigeration system, for the sake of reaching the optimal configuration of the whole system, but the placement of generation process of ARC and the selection of evaporation temperature were not well investigated. This work was later extended by taking Organic Rankine Cycle into consideration for process heat recovery in 2014 [12], but the ARC was still supposed to work at certain conditions rather than its optimal interior behavior accompanying with the system synthesis.

Although ARC is an effective technology, it cannot achieve a very low evaporation temperature [13], and the energy efficiency drops quickly with decreasing evaporation temperature and generation temperature [14,15]. By contrast, the electrical compression refrigeration (ECR) can reach much lower refrigerating temperature and usually has higher efficiency, whereas consuming expensive electricity [16]. Compression–absorption cascade refrigeration system (CACRS) retains the advantages of both ARC and ECR, as being able to achieve lower evaporation temperature with relatively high efficiency and less electricity consumption by recovering waste heat [17].

In last several years, the increasing researches on CACRS have been reported. Some studies investigated the optimal selection and combination of working fluids so that better performance can be achieved. Cimsit et al. [18] and Colorado et al. [19] made a thermodynamic analysis and comparison of the CACRS with various couples of working fluids in the absorption and vapor compression sections. According to the study, the highest performance was associated with the R134a–LiBr/H_2O based system. The earlier literature studies dealing with the performance improvement of CACRS are mainly investigated based on the first and second laws, such as Cimsit et al. [20]. Subsequently, the economic factor was taken into consideration in Cimsit's study [21], the exergy-based thermoeconomic analyses has been carried out and the best operating conditions can be determined with exergetic efficiency and minimum cost as objectives. The structural development of CACRS is also involved in relevant studies. Jain et al. [22] and Xu et al. [23] proposed the configurations of combined series-parallel and with evaporator-subcooler for cascade refrigeration system respectively and the energetic, exergetic, economic, and environmental analyses were conducted on different configurations.

In addition to above, some attention has been also concentrated to the optimal design of CACRS, in which parametric analysis on design variables are of great significance in pointing the direction for system design and optimization. According to Jain et al. [24], the change of operating temperatures contributes to significant changes in overall system performance. Jain et al. [25] established the model of CACRS, the effect of operating condition on the evaluation indicators are analyzed, based on which the study is extended to system optimization and design covering thermal and economical aspects [26]. Considering the conflict between maximizing exergetic efficiency and minimizing total cost rate, multi-objective optimization of CACRS is performed by Aminyavari et al. [27]. And the multi-objective optimization problem is solved with NSGA-II technique by Jain et al. [28].

As mentioned, previous researches only focused on the incorporation of ARC with HEN for heat recovery while without optimizing operating condition, and the coefficient of performance (COP) of ARC was also assumed to be constant. Considering that CACRS combines the advantages of ARC and ECR, it is meaningful to integrate HEN with CACRS, however no study has been launched. Therein, considering the coupling relationship between HEN and CACRS, the operating condition of CACRS will significantly affect the structure of HEN and the performance ($COPar_1$ for the absorption subsystem and $COPec_2$ for the compression subsystem are used as indicators to evaluate the performance) of CACRS. Therefore, in this paper, CACRS and HEN are considered as a whole system to recover heat for refrigeration, where the operating parameters and COPs of CACRS and the structure of HEN are optimized simultaneously.

The challenge of HEN synthesis with CACRS integrated is how to describe the coupling relationship between HEN and CACRS quantitatively, for which the COPs affected by operating condition are the crucial parameters to quantify the relationship. This paper handles the problem by means of discretizing the evaporation temperatures of two subsystems. At each evaporation temperature, the quantitative relationships between $COPar_1$ and heat source temperature, the quantitative relationships between $COPec_2$ and condensation temperature, are both obtained by simulating and data fitting. Therefore, the CACRS can be integrated into HEN in the form of a more detailed mathematic formulation. With this method, the optimal integrated system can be obtained with better performance and higher energy efficiency, large amount of excess heat can be recovered to minimize the economic cost.

This paper is organized as follows. The problem to be addressed in this paper is stated as Section 2. The simulation and model formulation are conducted in Section 3, wherein the two subsystems of CACRS are simulated, and the quantitative relationship between COPs and the operating variables of CACRS is determined with the simulation results, and the synthesis method for HEN integrated with CACRS is presented, which includes the superstructure composed of plentiful network alternatives and the detailed mathematical description for optimization purpose. The proposed method is demonstrated through two case studies in Section 4, and the discussion is presented in Section 5. Finally, conclusions are drawn in Section 6.

2. Problem Statement

Figure 1 shows the schematic representation of the CACRS synthesized with heat source. The system is composed of an absorption subsystem and a single stage electrical compression subsystem. These two subsystems share a heat exchanger which operates as the evaporator of absorption section meanwhile the condenser of compression section simultaneously. The generator of absorption subsystem is heated by heat sources either waste heat of hot process streams or external steam, and the evaporators of two subsystems provide process streams with cooling energy. The absorber and condenser are cooled down with cooling water. Here it should be noted that, when the required refrigerating temperature, lower than room temperature while no less than 0 °C, can be achieved by the evaporator of absorption subsystem, electrical compression subsystem is not needed any more. When the refrigerating temperature is too low to be achieved by ARC, below 0 °C, the demand of refrigeration is satisfied by compression subsystem's evaporation process of CACRS.

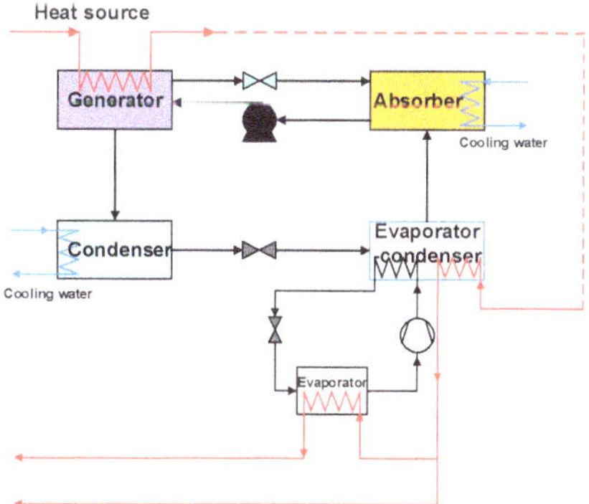

Figure 1. Schematic representation of the compression–absorption cascade refrigeration system (CACRS).

The COPs are crucial parameters to establish the relationship between the cold energy requirements in the evaporator and heat required by the generator (or the electricity consumed by the compressor). The mathematical expressions of $COPar_1$ for ARC and $COPec_2$ for ECR are in Equations (1) and (2), respectively. Whereas, $COPar_1$ and $COPec_2$ are reflected by the operating condition of ARC and ECR subsystems. Thus, in addition to exploring the structural integration of HEN and CACRS (or ARC or ECR if only one subsystem left), the COP related operational parameters, are also optimized with in this study.

$$COPar_1 = \frac{cooling\ output}{heat\ input} \tag{1}$$

$$COPec_2 = \frac{cooling\ output}{power\ input} \tag{2}$$

The synthesis problem to be addressed in this paper can be stated as follows.

A set of hot process streams (some are below room temperature or below 0 °C), and a set of cold process streams are given with their heat capacity flow rates, supply and target temperatures. Parameters of utilities, such as their inlet temperatures, outlet temperatures and unit costs are also given. Other parameters given include the cost coefficients of heat transfer units, the minimum approach temperatures and so on. Then, the problem consists in determining the optimal design of HEN coupled with CACRS including the optimal design of HEN structure and the operating condition of CACRS, as well as the trade-off between the efficiency and amount of heat recovery, with the aim of minimizing the total annual cost (TAC). Thus, to solve the problem, the relationship between COPs and operating condition is simulated in this paper and based on which, the mathematical model of HEN integrated with CACRS is established for optimal design.

3. Simulation and Model Formulation

3.1. Simulation of Compression–Absorption Cascade Refrigeration System

The CACRS is modelled and simulated with the process modelling software Aspen Plus. Two steady state simulation models for the two subsystems have been structured and implemented based on sequential modular approach in the computer program. The simulation used LiBr-H$_2$O fluid pairs within absorption subsystem and R134a within compression subsystem. The ELECNRTL and STEAMNBS property methods in Aspen Plus are chosen for LiBr-H$_2$O solution and pure water. The RK-SOAVE property method is used for R134a. It is assumed that the pressure and heat losses are neglectable, working fluid at the exits of generator, absorber, evaporator, cascaded evaporator-condenser and condenser are all in saturation state. The model is validated by comparing with the data presented by Cimsit et al. [21].

In this study, the condensation temperature is set as 45 °C and the concentration of solution is set as 0.58 for absorption section, referring to Wang's work [29]. A series of evaporation temperatures alternatives (5 °C, 7 °C, 9 °C, 11 °C, 13 °C, 15 °C) are given for choosing, and the heat source temperature will be varied in certain range (from 105 °C to 175 °C). For the compression section, the optional evaporation temperatures are optimized in a series of values (−25 °C, −21 °C, −17 °C, −13 °C, −9 °C, −5 °C) according to refrigeration demands, and the corresponding condensation temperature are 10 °C, 12 °C, 14 °C, 16 °C, 18 °C, and 20 °C when the temperature difference of cascaded evaporator-condenser is supposed to be 5 °C. Figure 2a,b present the variation of COP with heat source temperature and condensation temperature at certain evaporation temperature for absorption and compression section separately. Obviously, $COPar_1$ is improved with the increase of heat source temperature (t) and evaporation temperature. Higher evaporation temperature (te) and lower condensation temperature (tc) contribute to higher $COPec_2$. The corresponding functional relationship between $COPar_1$ and heat source temperature can be obtained in form of fitting curve, which is performed within MATLAB. The values calculated according to fitting function are compared with the simulated value to examine the accuracy, and the maximum error is less than 2%. In the same way, the matrix $COPec_2$ determined by evaporation and condensation temperatures can be also established.

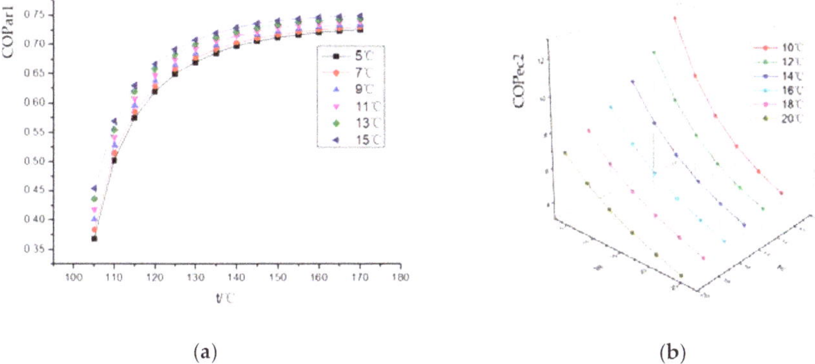

(**a**)　　　　　　　　　　　　　　　　　　(**b**)

Figure 2. (**a**) Effect of heat source temperature on coefficient of performance (COP) of absorption refrigeration cycle (ARC); (**b**) Effect of evaporation temperature and condensation temperature on COP of electrical compression refrigeration (ECR).

3.2. Superstructure Presentation

A stage-wise superstructure is proposed to present the configuration of HEN incorporated with the generation and evaporation processes of CACRS in Figure 3. The cooperation of HEN and CACRS can be performed in two parts: (i) Hot process streams provide heat to drive the generators of absorption subsystem at the inner stage, (ii) low-temperature cooling energy produced in evaporators of absorption and compression subsystems are used to cool hot process streams to sub-ambient temperature or below 0 °C at the stream ends after the use of cooling water.

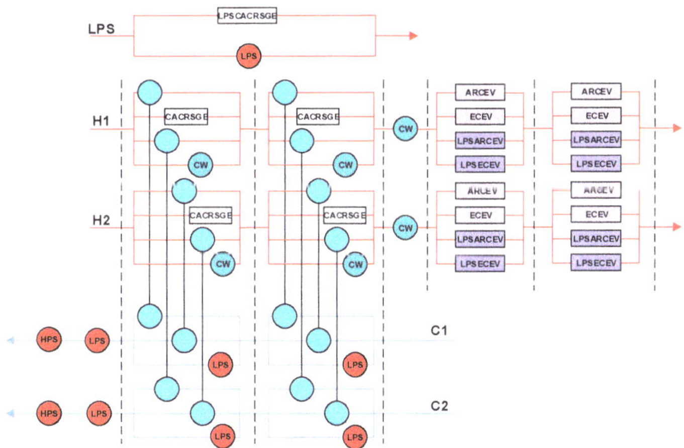

Figure 3. Superstructure of the heat exchanger network (HEN) integrated with CACRS.

As indicated, the generation (CACRSGE) and evaporation processes of absorption (ARCEV) and compression (ECREV) subsystems are implemented cooperatively with the heat exchange of mentioned process streams. Also, low-pressure steam (LPS) can be used as heat source to motivate CACRS for producing cooling energy when hot process streams can't provide enough heat. The corresponding generation and evaporation process are LPSCACRSGE, LPSARCEV, and LPSECREV. Each hot process stream can be cooled down by cold process stream, cooling water and/or the generation process (CACRSGE) and evaporation processes (ARCEV, ECREV, LPSARCEV, and LPSECREV) of CACRS, and each cold process stream can be heated by hot process stream, LPS and/or high-pressure steam (HPS).

In addition, a two-stage structure of heat exchange between hot process stream and the evaporation processes of CACRS are performed at the stream end in order to include more matching possibilities.

3.3. Model Formulation

A MINLP model consisting of HEN and CACRS is formulated according to the superstructure proposed in this paper. The following sets are defined before presenting the model formulation: *I*, *J*, *ST*, *ST'* represent the sets for hot process streams, cold process streams, stage numbers of heat exchange inside the superstructure, and stage numbers of heat exchange at end of the superstructure, *S* and *L* are the sets for temperature grades of evaporation processes for CACRS.

3.3.1. Model for Heat Exchanger Network Synthesis

- Total energy balances for process streams

$$\left(T_i^{in} - T_i^{out}\right)FCp_i = \sum_k \sum_j q_{i,j,k} + \sum_k q_{i,k}^{CACRSGE} + \sum_k q_{i,k}^{CW} + q_i^{CW'} + \sum_{k'} q_{i,k'}^{CACRSEV} \ \forall i \in I \tag{3}$$

$$\left(T_j^{out} - T_j^{in}\right)FCp_j = \sum_k \sum_i q_{i,j,k} + \sum_k q_{j,k}^{LPS} + q_j^{LPS'} + q_j^{HPS} \ \forall j \in J \tag{4}$$

Equations (3) and (4) represent the total energy balances for hot stream *i* and cold stream *j*. $q_{i,k}^{CACRSGE}$ denotes the heat load provided by hot process stream *i* in generator of CACRS within stage *k*, $q_{i,k'}^{CACRSEV}$ denotes the cooling energy provided by the evaporation processes of CACRS within stage *k'* at the stream end. $q_{i,k}^{CW}$ and $q_i^{CW'}$ denote the use of cooling water inner stage and at the stream end of the superstructure. $q_{j,k}^{LPS}$ and $q_j^{LPS'}$ denote the similar position of LPS as the placement of cooling water.

- Energy balance for each match inner stage of the superstructure

$$q_{i,j,k} = fcp_{i,j,k}\left(t_{i,j,k}^{in} - t_{i,j,k}^{out}\right) \ \forall i \in I, \forall j \in J, \forall k \in ST \tag{5}$$

$$q_{i,k}^{CACRSGE} = fcp_{i,k}^{CACRSGE}\left(t_{i,k}^{CACRSGEin} - t_{i,k}^{CACRSGEout}\right) \ \forall i \in I, \forall k \in ST \tag{6}$$

$$q_{i,k}^{CW} = fcp_{i,k}^{CW}\left(t_{i,k}^{CWin} - t_{i,k}^{CWout}\right) \ \forall i \in I, \forall k \in ST \tag{7}$$

$$q_{i,j,k} = fcp_{j,i,k}\left(t_{j,i,k}^{out} - t_{j,i,k}^{in}\right) \ \forall i \in I, \forall j \in J, \forall k \in ST \tag{8}$$

$$q_{j,k}^{LPS} = fcp_{j,k}^{LPS}\left(t_{j,k}^{LPSout} - t_{j,k}^{LPSin}\right) \ \forall j \in J, \forall k \in ST \tag{9}$$

It should be note that $t_{i,j,k}^{in}$, $t_{i,k}^{CACRSGEin}$, $t_{i,k}^{CWin}$ all equal to $t_{i,k}$, $t_{j,i,k}^{in}$ and $t_{j,k}^{LPSin}$ equal to $t_{j,k+1}$, on account that the inlet temperature of stage *k* is the outlet temperature after mixing in previous stage.

- Non-isothermal mixing for each inner stage

$$FCp_i = \sum_j fcp_{i,j,k} + fcp_{i,k}^{CACRSGE} + fcp_{i,k}^{CW} \ \forall i \in I, \forall k \in ST \tag{10}$$

$$\left(t_{i,k} - t_{i,k+1}\right)FCp_i = \sum_j q_{i,j,k} + q_{i,k}^{CACRSGE} + q_{i,k}^{CW} \ \forall i \in I, \forall k \in ST \tag{11}$$

$$FCp_j = \sum_i fcp_{j,i,k} + fcp_{j,k}^{LPS} \ \forall j \in J, \forall k \in ST \tag{12}$$

$$\left(t_{j,k} - t_{j,k+1}\right)FCp_j = \sum_i q_{i,j,k} + q_{j,k}^{LPS} \ \forall j \in J, \forall k \in ST \tag{13}$$

For process streams, the mass balances and energy balances for internal stages of the superstructure are shown as Equations (10)–(13). With these equations, the outlet temperature of stage k after mixing can be calculated.

- Energy balances for matches of evaporation processes at the stream ends

$$\left(t_{i,k'}^{CACRSEVin} - t_{i,k'}^{ARCEVout}\right)fcp_{i,k'}^{ARCEV} = q_{i,k'}^{ARCEV} \;\forall i \in I, \forall k' \in ST' \tag{14}$$

$$\left(t_{i,k'}^{CACRSEVin} - t_{i,k'}^{ECREVout}\right)fcp_{i,k'}^{ECREV} = q_{i,k'}^{ECREV} \;\forall i \in I, \forall k' \in ST' \tag{15}$$

$$\left(t_{i,k'}^{CACRSEVin} - t_{i,k'}^{LPSARCEVout}\right)fcp_{i,k'}^{LPSARCEV} = q_{i,k'}^{LPSARCEV} \;\forall i \in I, \forall k' \in ST' \tag{16}$$

$$\left(t_{i,k'}^{CACRSEVin} - t_{i,k'}^{LPSECREVout}\right)fcp_{i,k'}^{LPSECREV} = q_{i,k'}^{LPSECREV} \;\forall i \in I, \forall k' \in ST' \tag{17}$$

The heat removed by the evaporation processes of CACRS in any stage k' which is at the ends of hot streams are given by Equations (14)–(17). $q_{i,k'}^{ARCEV}$ and $q_{i,k'}^{LPSARCEV}$ denote the heat removed by the evaporation process of absorption section motivated by hot process streams and LPS at stage k' of hot process stream end, likewise, $q_{i,k'}^{ECREV}$ and $q_{i,k'}^{LPSECREV}$ have the similar meaning with $q_{i,k'}^{ARCEV}$ and $q_{i,k'}^{LPSARCEV}$, but the evaporation process belongs to the compression section of CACRS.

- Energy balance of non-isothermal mixing for evaporation processes at the stream ends

$$FCp_i = fcp_{i,k'}^{ARCEV} + fcp_{i,k'}^{ECREV} + fcp_{i,k'}^{LPSARCEV} + fcp_{i,k'}^{LPSECREV} \;\forall i \in I, \forall k' \in ST' \tag{18}$$

$$q_{i,k'}^{CACRSEV} = q_{i,k'}^{ARCEV} + q_{i,k'}^{ECREV} + q_{i,k'}^{LPSARCEV} + q_{i,k'}^{LPSECREV} \;\forall i \in I, \forall k' \in ST' \tag{19}$$

- Energy balances for utilities at the stream ends

$$\left(t_{i,NOK+1} - t_{i,1}^{CACRSin}\right)FCp_i = q_i^{CW'} \;\forall i \in I \tag{20}$$

$$\left(t_{i,k'}^{CACRSin} - t_{i,k'}^{CACRSout}\right)FCp_i = q_{i,k'}^{CACRSEV} \;\forall i \in I \tag{21}$$

$$\left(t_j^{HPSin} - t_{j,1}\right)FCp_j = q_j^{LPS'} \;\forall j \in J \tag{22}$$

$$\left(T_j^{out} - t_j^{HPSin}\right)FCp_j = q_j^{HPS} \;\forall j \in J \tag{23}$$

At the end of the superstructure, the heat removed by cooling water is determined by Equation (20), and the heat removed by the evaporation processes of CACRS are given by Equation (21). With Equations (22) and (23), the heating requirement of each cold process stream at the stream end are satisfied by LPS and HPS are determined.

- Feasibility constraints of temperature

$$t_{i,M}^{in} \geq t_{i,M}^{out} \;\forall i \in I \tag{24}$$

$$t_{j,N}^{in} \leq t_{j,N}^{out} \;\forall j \in J \tag{25}$$

$$T_i^{in} \geq t_{i,k} \geq t_{i,k+1} \geq T_i^{out} \;\forall i \in I \tag{26}$$

$$T_j^{out} \geq t_{j,k} \geq t_{j,k+1} \geq T_j^{in} \;\forall j \in J \tag{27}$$

Equation (24) denotes the temperature of hot process stream i will decrease after exchanging heat with M, M is the set including cold process stream, generator of CACRS, cooling water, and evaporators of CACRS. Similarly, Equation (25) denotes the temperature of cold process stream j will increase after exchanging heat with N which contains hot process stream, LPS and HPS. Equations (26) and (27) are used to ensure the monotonic decrease of temperature from the left side to the right side of the superstructure.

- Constraints on binary variables of CACRS

$$\sum_s Z^{CACRSGE}_{i,k,s} \leq 1 \ \forall i \in I, \forall k \in ST \tag{28}$$

$$\sum_s Z^{ARCEV}_{i,k',s} \leq 1 \ \forall i \in I, \forall k' \in ST' \tag{29}$$

$$\sum_s \sum_l Z^{ECREV}_{i,k',s,l} \leq 1 \ \forall i \in I, \forall k' \in ST' \tag{30}$$

$$\sum_s Z^{LPSARCEV}_{i,k',s} \leq 1 \ \forall i \in I, \forall k' \in ST' \tag{31}$$

$$\sum_s \sum_l Z^{LPSECREV}_{i,k',s,l} \leq 1 \ \forall i \in I, \forall k' \in ST' \tag{32}$$

$Z^{CACRSGE}_{i,k,s}$ is the binary variable denoting the existence of generator of CACRS in stage k with the temperature of cooling energy produced in the evaporator belongs to grade s. $Z^{ARCEV}_{i,k',s}$ and $Z^{LPSARCEV}_{i,k',s}$ are binary variables that denote the cooling energy produced in grade s by absorption subsystem is used to cool hot process stream i at stage k'. $Z^{ECREV}_{i,k',s,l}$ and $Z^{LPSECREV}_{i,k',s,l}$ are binary variables which denote hot process streams are cooled down by the evaporation processes of CAVRS's compression section in stage k', of which s and l are the evaporation temperature grades of absorption section and compression section respectively.

The constraints on binary variable of CACRS are given by Equations (28)–(32). Equation (28) means that for the generation process of CACRS, there is at most one corresponding evaporation temperature of certain grade. Equations (29)–(32) indicate that for evaporation processes of CACRS, only one grade of evaporation temperature is considered so as to simplify the model and shorten computing time.

3.3.2. Model for CACRS

In the CACRS, condensation temperature and absorption temperature can be regarded as constants since the absorber and condenser are cooled down with cooling water which is used as cold utility in HEN with given inlet and outlet temperatures.

- The determination of matrix COP

$$copar_1(s) = f(t^{CACRSGEin}_{i,k}) \ COPec_2 = \begin{pmatrix} cop^{ec2}_{s_1,l_1} & \cdots & cop^{ec2}_{s_1,L} \\ \vdots & cop^{ec2}_{s,l} & \vdots \\ cop^{ec2}_{S,l_1} & \cdots & cop^{ec2}_{S,L} \end{pmatrix} \tag{33}$$

In the case that condensation and absorption temperatures of absorption section are defined, $COPar_1$ is only related to the inlet temperature of generator (heat source temperature) $t^{CACRSGEin}_{i,k}$ and the evaporation temperature T^{ARCEV} for ARC, $COPec_2$ can be determined by the condenser temperature (equal to $T^{ARCEV} + \Delta T_{min2}$) and evaporation temperature T^{ECREV} for ECR. Evaporation temperature of two subsystems are graded by discretizing into several values. Once the grades of these two evaporation temperatures are determined, $copar_1(s)$ can be expressed as the function of $t^{CACRSGEin}_{i,k}$ and the matrix of $COPec_2$ can be established. The functional relation and values in matrix are obtained through process simulation and data fitting which are presented in Section 3.1.

- Energy balance for CACRS

$$\sum_i \sum_k q^{CACRSGE}_{i,k} Z^{CACRSGE}_{i,k,s} \times copar_1(s) = \sum_i \sum_{k'} (Z^{ARCEV}_{i,k',s} q^{ARCEV}_{i,k'}) + \sum_i \sum_l \sum_{k'} (Z^{ECREV}_{i,k',s,l} q^{ECREV}_{i,k'}) \times (1 + 1/copec_2(s,l)) \tag{34}$$

$$q^{LPSCACRSGE}_s \times copar_1(s) = \sum_i \sum_{k'} (Z^{LPSARCEV}_{i,k',s} q^{LPSARCEV}_{i,k'}) + \sum_i \sum_{k'} \sum_l (Z^{LPSECREV}_{i,k',s,l} q^{LPSECREV}_{i,k'}) \times (1 + 1/copec_2(s,l)) \tag{35}$$

$q_s^{LPSCACRSGE}$ denotes the heat input from LPS to generator. Equations (34) and (35) give the energy balances of the CACRS, which means that the evaporator of absorption subsystem get heat from hot process stream and the condenser of compression subsystem.

3.3.3. Objective Function

The heat exchanging area can be calculated with Equation (36), herein, h_M and h_N represent the film heat transfer coefficients, dt_1 and dt_2 are the temperature differences before and after heat exchanging, the mean temperature difference is approximated according to Chen [30]. Objective of the optimization is to minimize the TAC, considering both the expenditures of capital and operation for HEN and CACRS. The corresponding cost representations are given in Equations (38)–(40). HEN cost is the summation of capital cost for heat exchangers and operating cost for utility consumption. CACRS cost is deduced from the investment of absorption subsystem (containing generator, evaporator, condenser and absorber) and compression subsystem (cascaded evaporation-condenser, compressor and evaporator are included), as well as the cooling water consumption in the absorber and condenser which belong to absorption subsystem and the electricity consumption of compressor in compression subsystem.

$$A = \frac{q \times \left(\frac{1}{h_M} + \frac{1}{h_N}\right)}{\left(dt_1 \times dt_2 \times \frac{dt_1 + dt_2}{2}\right)^{\frac{1}{3}}} \tag{36}$$

$$Object\ function = minTAC = TAC^{HEN} + TAC^{CACRS} \tag{37}$$

$$TAC^{HEN} = \left\{ \begin{aligned} & K_f \times \alpha_{HEN}^{fixed} \times \left(\sum_i \sum_j \sum_k Z_{i,j,k} + \sum_i \sum_k Z_{i,k}^{CW} + \sum_i Z_i^{CW'} + \sum_j \sum_k Z_{j,k}^{LPS} + \sum_j (Z_j^{LPS'} + Z_j^{HPS})\right) \\ & + K_f \times \alpha_{HEN}^{area} \times \left(\sum_i \sum_j \sum_k (A_{i,j,k})^{\beta_{HEN}^{area}} + \sum_i \sum_k (A_{i,k}^{CW})^{\beta_{HEN}^{area}}\right. \\ & + \sum_i (A_i^{CW'})^{\beta_{HEN}^{area}} + \sum_j \sum_k (A_{j,k}^{LPS})^{\beta_{HEN}^{area}} + \sum_j \left((A_j^{LPS'})^{\beta_{HEN}^{area}} + (A_j^{HPS})^{\beta_{HEN}^{area}}\right) \\ & + C_{cw} \times \left(\sum_i q_i^{CW'} + \sum_i \sum_k q_{i,k}^{CW}\right) + C_{hps} \times \sum_j q_j^{HPS} + C_{lps} \times \left(\sum_j \sum_k q_{j,k}^{LPS} + \sum_j q_j^{LPS'}\right) \end{aligned} \right\} \tag{38}$$

$$TAC^{CACRS} = \left\{ \begin{aligned} & K_f \times \alpha_{CACRS}^{fixed} \times \left(\sum_i \sum_k \sum_s Z_{i,k,s}^{CACRSGE} + \sum_i \sum_{k'} \sum_s Z_{i,k',s}^{ARCEV} + \sum_i \sum_{k'} \sum_s \sum_l Z_{i,k',s,l}^{ECREV} + Z^{CACRSCON}\right. \\ & + Z^{CACRSABS} + \sum_s Z_s^{LPSCACRSGE} + \sum_i \sum_{k'} \sum_s Z_{i,k',s}^{LPSARCEV} + \sum_i \sum_{k'} \sum_s \sum_l Z_{i,k',s,l}^{LPSECREV}\Big) + Z^{LPSCACRSCON} \\ & + Z^{LPSCACRSABS} K_f \times \alpha_{CACRS}^{area} \times \left(\sum_i \sum_k \sum_s A_{i,k,s}^{CACRSGE} + \sum_i \sum_{k'} \sum_s A_{i,k',s}^{ARCEV} + \sum_i \sum_k \sum_s \sum_l A_{i,k',s,l}^{ECREV}\right. \\ & + A^{CACRSCON} + A^{CACRSABS} + \sum_s A_s^{LPSCACRSGE} + + \sum_i \sum_{k'} \sum_s A_{i,k',s}^{LPSARCEV} + \sum_i \sum_k \sum_s \sum_l A_{i,k',s,l}^{LPSECREV} \\ & + A_{absorber}\Big) + A^{LPSCACRSCON} + A^{LPSCACRSABS} + C_{comp} + \left(C_{cw} \times (q_{condenser} + q_{absorber})\right. \\ & + C_{lps} \times \sum_s q_s^{LPSCACRSGE} + H_Y \times C_{elec} \times Wele) \end{aligned} \right\} \tag{39}$$

4. Case Study

In this section, the proposed model is demonstrated through two case studies. To make comparison, a base case is set to cool down process streams by employing electrical compression refrigeration (R134a), which produces cooling energy by consuming electricity rather than process heat. Analogously, for the HEN integrated with ECR, each hot process stream is allowed to be cooled down by multiple parallel evaporation processes belonging to different grades after the using of cooling water. In the absorption subsystem of CACRS, LiBr/H_2O is used as the working fluid. For single LiBr/H_2O absorption refrigeration, the temperature of heat source is not allowed to be lower than 80 °C or higher than 180 °C. In this paper, the $COPar_1$ lower than 0.5 is not considered, so the designated range of heat source temperature is 105 °C to 175 °C. The minimum approach temperature for process stream matching and CACRS matching are stipulated to be $\Delta T_{min1} = 10$ °C, $\Delta T_{min2} = 5$ °C respectively. The inlet and outlet temperatures of cooling water are 30 °C and 40 °C, temperatures of LPS and HPS are

158 °C and 255 °C respectively. Similar to CACRS, the condenser of ECR is also cooled down by cooling water, and accordingly, the COP of ECR (*COPec*) can be determined by the selectable evaporation temperatures $T^{ECREV'}$ (denoted by s and l) which are identical to CACRS.

The data of the involved process streams for the two cases are shown in Tables 1 and 2. In case 2, one more hot process stream is added into the HEN, which means more heat available, meanwhile more demand for cooling.

Table 1. Stream data for case 1.

Streams	Inlet Temperature $T_{in}/°C$	Outlet Temperature $T_{out}/°C$	$FCp/kW \bullet °C^{-1}$
H1	135	−5	38.75
H2	145	20	48.75
C1	35	240	20
C2	60	120	36.25

Table 2. Stream data for case 2.

Streams	Inlet Temperature $T_{in}/°C$	Outlet Temperature $T_{out}/°C$	$FCp/kW \bullet °C^{-1}$
H1	165	−15	42.5
H2	135	−5	38.75
H3	145	20	48.75
C1	35	240	20
C2	60	120	36.25

In order to optimize the operating condition of CACRS, six evaporation temperature alternatives for absorption subsystem and compression subsystem are given and graded respectively according to the requirement of target temperature. The grading about evaporation temperatures for absorption subsystem is shown in Table 3. Similarly, the evaporation temperature for compression subsystem is also graded into six levels in the range of −25 °C to −5 °C, with l_1 to l_6 denoting −25 °C, −21 °C, −17 °C, −13 °C, −9 °C, and −5 °C respectively. Based on the simulation and data fitting procedures stated in Section 3, the functional relation of $COPar_1$ with $t_{i,k}^{CACRSGEin}$ is obtained and presented as Equation (40), and the correlation coefficients are listed in Table 3. Corresponding $cop_{s,l}^{ec2}$ in matrix $COPec_2$ are given in Equation (41) according to the general presentation of Equation (33). When only ECR is used to satisfy the refrigeration demand, the $cop_{s,l}^{ec}$ in matrix $COPec$ is given in Equation (42), wherein, the cop_s^{ec} used to produce cooling energy below room temperature is given as Equation (43), and the cop_l^{ec} used to produce cooling energy lower than 0 °C is given as Equation (44).

$$COPar_1(s) = a \times 10^{-6} \times (t_{i,k}^{CACRSGEin})^3 + b \times 10^{-3} \times (t_{i,k}^{CACRSGEin})^2 + c \times 10^{-1} \times (t_{i,k}^{CACRSGEin}) + d \quad (40)$$

$$COPec_2 = \begin{pmatrix} 4.64 & 5.41 & 6.42 & 7.77 & 9.7 & 12.66 \\ 4.34 & 5.03 & 5.91 & 7.08 & 8.69 & 11.06 \\ 4.07 & 4.69 & 5.47 & 6.49 & 7.86 & 9.8 \\ 3.83 & 4.39 & 5.08 & 5.97 & 7.15 & 8.77 \\ 3.61 & 4.11 & 4.74 & 5.53 & 6.55 & 7.93 \\ 3.40 & 3.86 & 4.43 & 5.13 & 6.03 & 7.22 \end{pmatrix} \quad (41)$$

$$COPec(s/l) = \begin{pmatrix} cop_{s_1/l_1}^{ec} & \cdots & cop_{s/l}^{ec} & \cdots & cop_{S/L}^{ec} \end{pmatrix} \quad (42)$$

$$COPec(s) = \begin{pmatrix} 4.13 & 4.43 & 4.75 & 5.12 & 5.53 & 6 \end{pmatrix} \quad (43)$$

$$COPec(l) = \begin{pmatrix} 1.78 & 1.96 & 2.17 & 2.42 & 2.70 & 3.02 \end{pmatrix} \quad (44)$$

Table 3. Coefficients of the function relation of $COPar_1$ with $t_{i,k}^{CACRSGEin}$.

s	$T^{ARCEV}/°C$	a	b	c	d
s_1	5	3.5075	−1.5798	2.3717	−11.158
s_2	7	3.6281	−1.6257	2.4273	11.362
s_3	9	3.5897	−1.6077	2.3976	11.186
s_4	11	3.2346	−1.4532	2.1756	10.127
s_5	13	3.0330	−1.3649	2.0456	9.4823

The formulated MINLP mathematical models for mentioned HEN-CACRS and HEN-ECR designs are coded and solved in GAMS with PC specification: CPU 3.30 GHz, 8 GB RAM. For each model, BARON is used as solver. To be clear, note that the global optimum cannot be guaranteed for MINLP problems. Yet, good initial values and boundaries as well as circular iteration are employed to facilitate the search process and improve the quality of locally-optimal solutions.

In HEN-CACRS result, the evaporation temperature of grade 's' produced in ARC subsystem can cool hot process stream to below room temperature, and the evaporation temperature of grade 'l', produced in ECR subsystem, can be used for cooling hot process stream to below 0 °C. For the integrated CACRS, (s, l) denotes that there is cooling energy produced at evaporation temperature of grade 'l' by the compression section, whose temperature of cascaded evaporator-condenser belongs to grade 's', while for HEN-ECR result, just the evaporation temperature grade of 's' or 'l' needs to be determined. Optimal structures of the integrated HENs with CACRS and ECR for two cases are presented in Figure 4a,b and Figure 5a,b, respectively.

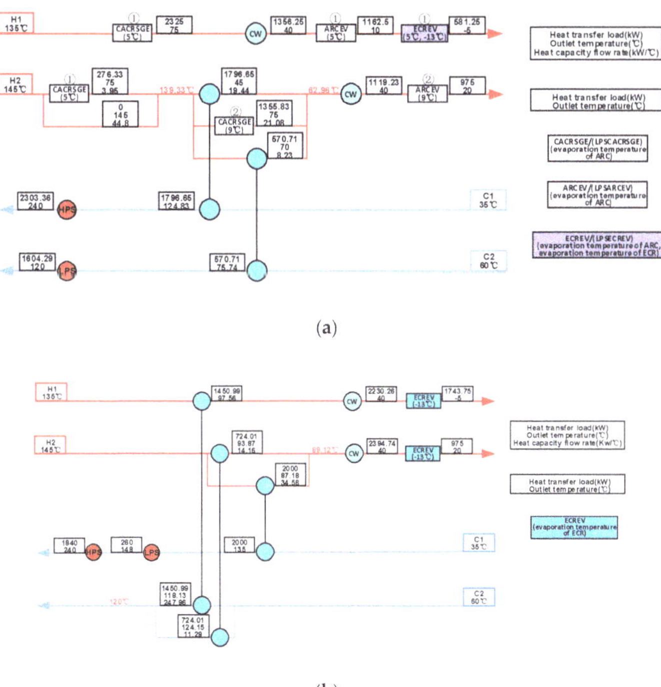

(a)

(b)

Figure 4. Result of case 1. (**a**) Structure of the integrated HEN with CACRS; (**b**) Structure of the integrated HEN with ECR.

Figure 4a shows the optimal structure of HEN integrated with CACRS for case 1, as indicated in which one set of CACRS and one set of ARC are implemented to absorb heat from process stream for refrigeration purpose. The CACRSGE (5 °C) is motivated by H1 and H2, and the cooling energy, 5 °C (s_1) produced by the evaporation process of absorption section, and −13 °C (l_4) produced in evaporator of compression section are used to cool down H1 (ARCEV(5 °C), and ECREV (5 °C, −13 °C)) successively. In addition to exchanging heat with cold process stream and the generator of CACRS, H2 heats the generator of the ARC (CACRSGE (9 °C)), then to be refrigerated by the evaporation process of ARC (ARCEV(9 °C), (s_3)) after the use of cooling water. In HEN-ECR result, as shown in Figure 4b, the mentioned refrigeration demands are all satisfied by ECRs. The evaporating temperature for refrigerating H1 and H2 are −13 °C (l_4).

For case 2, the optimal integrated structures of the two concerned design scenarios are shown in Figure 5a,b.

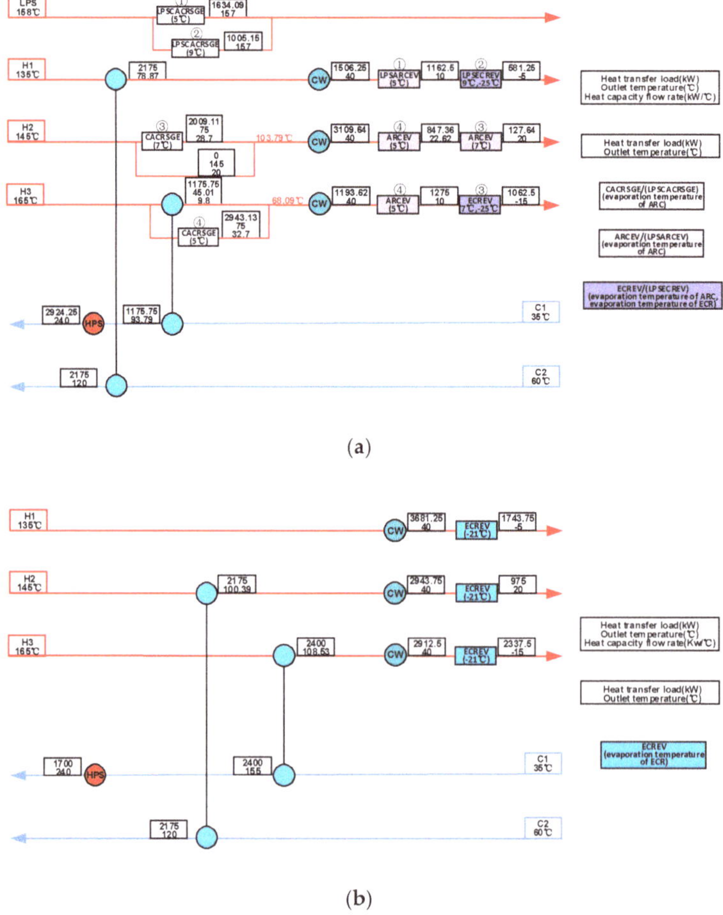

(a)

(b)

Figure 5. Result of case 2. (**a**) Structure of the integrated HEN with CACRS; (**b**) Structure of the integrated HEN with ECR.

For the scenario HEN integrated with CACRS, two ARC sets and two CACRS sets are involved to provide cooling energy to HEN. Since the heat of process streams is not sufficient, LPS is used to compensate the heat demand of CACRS. One ARC set is run by LPS (LPSCACRSGE (5 °C)), and the evaporator absorbs heat from H1 (LPSARCEV (5 °C), (s_1)). The other ARC set gets heat from H3 in generator (CACRSGE (5 °C)) and the cooling energy produced in the evaporator is used to cool H2 and H3 (ARCEV (5 °C), (s_1)). One CACRS is motivated by LPS (LPSCACRSGE (9 °C)), and the cooling energy produced in the evaporator of compression section is used to cool down H1 (LPSECREV (9 °C, −25 °C), (s_3, l_1)) at the stream end. The other CACRS gets heat from H2 (CACRSGE (7 °C)) and the cooling energy produced in the evaporator of absorption and compression subsystems is used for cooling H2 (ARCEV (7 °C), (s_2)) and H3 (ECREV (7 °C, −25 °C), (s_2, l_1)) to their target temperatures separately.

5. Discussion

From obtained results of the two cases, it can be observed that, in HEN-CACRS and HEN-ECR scenarios, the process streams with cooling requirements are always cooled to the lowest temperature which can be achieved by cooling water first, and then are refrigerated by the evaporation processes of refrigeration system subsequently. The cause of this state is that cooling energy is produced at the cost of consuming much heat and electricity which will increase the operating cost. What is more, by comparing the optimal structure of HEN-CACRS with HEN-ECR, it can be observed that HEN-ECR prefers equal or higher evaporating temperature grades than that of CACRS at the last stage of evaporation processes. The reason for this tendency is that, higher evaporating temperature is beneficial for higher COP, which will decrease the consumption of expensive electricity.

Higher COP is pursued by both ECR and CACRS, which means consuming less electricity and heat. But when CACRS is integrated into HEN, high COP cannot be reached all the time due to the limit of cost target and characteristics of process streams (inlet temperature, outlet temperature, heat load and so on), there are trade-offs between the quantity and efficiency of heat recovery and between the cost of hot utility and electricity consumption.

For the absorption section, the lower the evaporation temperatures, the higher the hot source temperature (inlet temperature of generator) will be needed for higher $COPar_1$. Such as the evaporation temperature used to cool H2 in case1 is 9 °C (s_3), the corresponding heat source temperature is 139.33 °C, the evaporating temperature which is used to cool H1 is 5 °C (s_1), and the matching heat temperature is 145 °C. But there is also 5 °C (s_1) cooling energy produced, which is motivated the heat source of 135°C, for recovering much heat. Likewise, for the compression subsystem of CACRS, in case 2, there are two matches of evaporation temperatures for ARC section and ECR section, (7 °C, −25 °C) and (9 °C, −25 °C), obviously, the match of (7 °C, −25 °C) has higher $COPec_2$, but the match of (9 °C, −25 °C) is motivated by LPS, higher evaporation temperature of ARC section will lead to higher $COPar_1$ and lower $COPec_2$ at the same time, so there is a trade-off between LPS and electricity consumption costs.

For the two cases, the optimal integrated HEN-ECR system are achieved at the TAC of 2,595,319 $\$\cdot y^{-1}$ and 4,780,274 $\$\cdot y^{-1}$, and the TAC of optimal integrated HEN-CACRS are 1,592,527 $\$\cdot y^{-1}$ and 2,970,188 $\$\cdot y^{-1}$, which are 38.6% and 37.9% lower than HEN-ECR respectively. The concrete analysis is as follows. The TAC consists of the capital cost of HEN, capital cost of CACRS and operating cost of HEN-CACRS. The operating cost of HEN-CACRS is composed of electricity cost, hot utility cost (LPS and HPS) and cold utility cost (cooling water). The cost of each part can be seen in Figure 6.

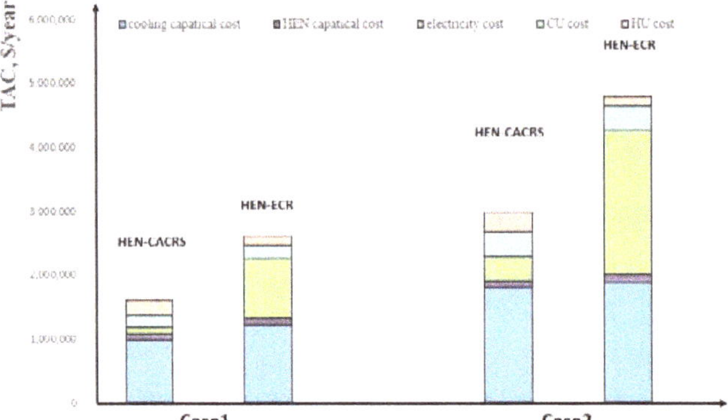

Figure 6. The distribution of total annual cost.

Apparently, the capital cost of refrigeration forms a considerable part of TAC, and the cost of CACRS is close to that of ECR for the two cases studied. By comparing the cost difference of HEN-CACRS with HEN-ECR, it can be found that there is little difference in the capital cost of HEN, the hot and cold utilities costs of HEN-CACRS are higher than that of HEN-ECR, but the electricity cost of HEN-CACRS is much less than that of HEN-ECR. There are the detail explanations. In HEN-CACRS scenario, the integration of CACRS helps to promote energy efficiency of hot process streams, some waste heat is used to motivate the generator of CACRS, and the energy produced by the evaporators of CACRS can be used to partly meet the refrigeration demand, which lead to sharp decrease of electricity cost. Even though the utilization of waste heat can decrease the consumption of cooling water for cooling hot process streams, the cooling water cost of HEN-CACRS is still a little higher than that of HEN-ECR, because the cooling water consumed by the refrigeration system is also considered. In addition, the integration of CACRS into HEN may lessen the heat load matching with cold process stream or increase the consumption of LPS as heat source to motivate the refrigeration system, and result in the increase of hot utility cost. But with comprehensive consideration of all costs, it can be concluded that the TAC of HEN-CACRS is still lower than that of HEN-ECR.

6. Conclusions

This paper has presented a mathematical programming model to consider the integration of HEN with CACRS. The coupling relationship of CACRS and HEN are quantitatively described through process simulating and data fitting. Then a detailed mathematical description for CACRS and HEN is formulated. Considering the mutual effect of HEN and CACRS, the structure of the HEN and the performance of CACRS are optimized simultaneously within the design. Compared with the results of base case where only electrical compression refrigeration is used for refrigeration purpose, it can be found that the integrated HEN-CACRS system is capable to utilize energy reasonably and reduce the total annualized cost by 38.6% and 37.9% respectively since it could recover waste heat from hot process stream to produce the cooling energy required by the system. On base of this study, future work will be concentrated on the optimal design of the HEN integrated with CACRS and ORC, and more internal parameters will be considered.

Author Contributions: Formal analysis, L.L.; Funding acquisition, J.D.; Investigation, Y.Z.; Methodology, X.S.; Project administration, J.D.; Software, L.Z.; Validation, L.L. and Y.Z.; Writing—original draft, X.S. and L.L.; Writing—review and editing, X.S. All authors have read and agreed to the published version of the manuscript.

Funding: This research is supported by Natural Science Foundation of China (Grant No. 21878034, 21776035), China Postdoctoral Science Foundation (2019TQ0045) and the Fundamental Research Fund for Central Universities of China (DUT18LAB11).

Conflicts of Interest: The authors declare no conflict of interest.

Nomenclature

Sets

I	set of hot process streams
J	set of cold process streams
ST	set of stages inner the superstructure
ST'	set of stages at end of the superstructure
S/L	set of evaporation temperature grades
M	set of process or processes need to be heated
N	set of process or processes need to be cooled

Parameters

T_i^{in}/T_i^{out}	inlet/outlet temperature of hot process streams, $^\circ$C
T_j^{in}/T_j^{out}	inlet/outlet temperature of cold process streams, $^\circ$C
FCp_i/FCp_j	heat capacity flowrate of hot/ cold process streams, kW$\cdot^\circ$C^{-1}
K_f	annualized factor, kg$\cdot$(m^3)$^{-1}$
$\alpha_{HEN}^{fixed}/\alpha_{HEN}^{area}/\beta_{HEN}^{area}$	cost parameters of heat exchanger network
$\alpha_{CACRS}^{fixed}/\alpha_{CACRS}^{area}$	cost parameters of cascade refrigeration system
H_Y	annual working hour, h
C_{cw}	unit cost of cooling water, \$/(kW$\cdot$y)
C_{lps}	unit cost of low-pressure steam, \$/(kW$\cdot$y)
C_{hps}	unit cost of high-pressure steam, \$/(kW$\cdot$y)
C_{elec}	unit cost of electricity, \$/(kW$\cdot$y)
C_{comp}	capital cost of compressor, \$$\cdot$y^{-1}

Superscripts

CACRS	compression-absorption cascade refrigeration system
LPS/LPS'	low-pressure steam
LPSCACRSGE	the evaporation process belongs to absorption subsystem of CACRS motivated by low-pressure steam
CACRSGE	the generation process of CACRS motivated by hot process stream
ARCEV	the evaporation process belongs to absorption subsystem of CACRS motivated by hot process stream
ECREV	the evaporation process belongs to compression subsystem of CACRS motivated by hot process stream
LPSARCEV	the evaporation process belongs to absorption subsystem of CACRS motivated by low-pressure steam
LPSECREV	the evaporation process belongs to compression subsystem of CACRS motivated by low pressure steam
CW/CW'	cooling water
HPS	high-pressure steam

Subscripts

i	hot process stream
j	cold process stream
k	index for inner stage of HEN superstructure (1, ... , NOK)
NOK	total number of inner stages
k'	index for evaporation stage at hot stream end
s	evaporation temperature grades of the CACRS's absorption subsystem
l	evaporation temperature grades of the CACRS's compression subsystem

Variables

t^{in}	inlet temperature, $^\circ$C
t^{out}	outlet temperature, $^\circ$C

fcp	heat capacity flowrate for a stream branch, $kW \cdot {}^{\circ}C^{-1}$
q	heat exchange load, kW
A	heat exchange area, m^2
Wele	the consumption of electricity, \$
$COPar_1$	coefficient of performance for the absorption section of CACRS
$COPec_2$	coefficient of performance for the compression section of CACRS
h_M/h_N	film heat transfer coefficient of the stream needs to be heated or cooled, $kW \cdot ({}^{\circ}C \cdot m^2)^{-1}$
dt_1/dt_2	temperature difference before or after heat exchanging, ${}^{\circ}C$

References

1. Alkanhal, T.A.; Sheikholeslami, M.; Usman, M.; Haq, R.; Shafee, A.; Al-Ahmadi, A.S.; Tlili, I. Thermal management of MHD nanofluid within the porous medium enclosed in a wavy shaped cavity with square obstacle in the presence of radiation heat source. *Int. J. Heat Mass Transf.* **2019**, *139*, 87–94. [CrossRef]

2. Sheikholeslami, M.; Jafaryar, M.; Shafee, A.; Li, Z.; Haq, R. Heat transfer of nanoparticles employing innovative turbulator considering entropy generation. *Int. J. Heat Mass Transf.* **2019**, *136*, 1233–1240. [CrossRef]

3. Yee, T.F.; Grossmann, I.E. Simultaneous optimization models for heat integration—II. Heat exchanger network synthesis. *Comput. Chem. Eng.* **1990**, *14*, 1165. [CrossRef]

4. Björk, K.M.; Westerlund, T. Global optimization of heat exchanger network synthesis problems with and without the isothermal mixing assumption. *Comput. Chem. Eng.* **2002**, *26*, 1581–1593. [CrossRef]

5. Pavão, L.V.; Costa, C.B.B.; Ravagnani, M.A.S.S. An enhanced stage-wise superstructure for heat exchanger networks synthesis with new options for heaters and coolers placement. *Ind. Eng. Chem. Res.* **2018**, *57*, 2560–2573. [CrossRef]

6. Ponce-Ortega, J.M.; Serna-González, M.; Jiménez-Gutiérrez, A. Synthesis of heat exchanger networks with optimal placement of multiple utilities. *Ind. Eng. Chem. Res.* **2010**, *49*, 2849–2856. [CrossRef]

7. Na, J.; Jung, J.; Park, C.; Han, C. Simultaneous synthesis of a heat exchanger network with multiple utilities using utility substages. *Comput. Chem. Eng.* **2015**, *79*, 70–79. [CrossRef]

8. Zhang, X.; Cai, L.; Chen, T. Energetic and exergetic investigations of hybrid configurations in an absorption refrigeration chiller by aspen plus. *Processes* **2019**, *7*, 609. [CrossRef]

9. Mussati, S.; Mansouri, S.; Gernaey, K.; Morosuk, T.; Mussati, M. Model-based cost optimization of double-effect water-lithium bromide absorption refrigeration systems. *Processes* **2019**, *7*, 50. [CrossRef]

10. Ponce-Ortega, J.M.; Tora, E.A.; González-Campos, J.B.; El-Halwagi, M.M. Integration of renewable energy with industrial absorption refrigeration systems: Systematic design and operation with technical, economic and environmental objectives. *Ind. Eng. Chem. Res.* **2011**, *50*, 9667–9684. [CrossRef]

11. Lira-Barragán, L.F.; Ponce-Ortega, J.M.; Serna-González, M.; El-Halwagi, M.M. Synthesis of integrated absorption refrigeration systems involving economic and environmental objectives and quantifying social benefits. *Appl. Therm. Eng.* **2013**, *52*, 402–419. [CrossRef]

12. Lira-Barragán, L.F.; Ponce-Ortega, J.M.; Serna-González, M.; El-Halwagi, M.M. Sustainable integration of trigeneration systems with heat exchanger networks. *Ind. Eng. Chem. Res.* **2014**, *53*, 2732–2750. [CrossRef]

13. Yan, X.; Chen, G.; Hong, D.; Lin, S.; Tang, L. A novel absorption refrigeration cycle for heat sources with large temperature change. *Appl. Therm. Eng.* **2013**, *52*, 179–186. [CrossRef]

14. Saleh, A.; Mosa, M. Optimization study of a single-effect water-lithium bromide absorption refrigeration system powered by flat-plate collector in hot regions. *Energy Convers. Manag.* **2014**, *87*, 29–36. [CrossRef]

15. Gogoi, T.K.; Konwar, D. Exergy analysis of a H_2O-LiCl absorption refrigeration system with operating temperatures estimated through inverse analysis. *Energy Convers. Manag.* **2016**, *110*, 436–447. [CrossRef]

16. Xu, Y.; Chen, F.S.; Wang, Q.; Han, X.; Li, D.; Chen, G. A novel low-temperature absorption—Compression cascade refrigeration system. *Appl. Therm. Eng.* **2015**, *75*, 504–512. [CrossRef]

17. Xu, Y.; Chen, G.; Wang, Q.; Han, X.; Jiang, N.; Deng, S. Performance study on a low-temperature absorption-compression cascade refrigeration system driven by low-grade heat. *Energy Convers. Manag.* **2016**, *119*, 379–388. [CrossRef]

18. Cimsit, C.; Ozturk, I.T. Analysis of compression-absorption cascade refrigeration cycles. *Appl. Therm. Eng.* **2012**, *40*, 311–317. [CrossRef]
19. Colorado, D.; Rivera, W. Performance comparison between a conventional vapor compression and compression-absorption single-stage and double-stage systems used for refrigeration. *Appl. Therm. Eng.* **2015**, *87*, 273–285. [CrossRef]
20. Cimsit, C.; Ozturk, I.T.; Hosoz, M. Second law based thermodynamic analysis of compression-absorption cascade refrigeration cycles. *J. Therm. Sci. Technol.* **2014**, *34*, 9–18.
21. Cimsit, C.; Ozturk, I.T.; Kincay, O. Thermoeconomic optimization of LiBr/H_2O-R134a compression-absorption cascade refrigeration cycle. *Appl. Therm. Eng.* **2015**, *76*, 105–115. [CrossRef]
22. Jain, V.; Sachdeva, G.; Kachhwaha, S.S. Energy, exergy, economic and environmental (4E) analyses based comparative performance study and optimization of vapor compression-absorption integrated refrigeration system. *Energy* **2015**, *91*, 816–832. [CrossRef]
23. Xu, Y.; Jiang, N.; Pan, F.; Wang, Q.; Gao, Z.; Chen, G. Comparative study on two low-grade heat driven absorption-compression refrigeration cycles based on energy, exergy, economic and environmental (4E) analyses. *Energy Convers. Manag.* **2016**, *133*, 535–547. [CrossRef]
24. Jain, V.; Kachhwaha, S.S.; Sachdeva, G. Thermodynamic performance analysis of a vapor compression–absorption cascaded refrigeration system. *Energy Convers. Manag.* **2013**, *75*, 685–700. [CrossRef]
25. Jain, V.; Sachdeva, G.; Kachhwaha, S.S. Thermodynamic modelling and parametric study of a low temperature vapour-compression absorption system based on modified Gouy-Stodola equation. *Energy* **2015**, *79*, 407–418. [CrossRef]
26. Jain, V.; Sachdeva, G.; Kachhwaha, S.S. NLP model based thermoeconomic optimization of vapor compression–absorption cascaded refrigeration system. *Energy Convers. Manag.* **2015**, *93*, 49–62. [CrossRef]
27. Aminyavari, M.; Najafi, B.; Shirazi, A.; Rinaldi, F. Exergetic, economic and environmental (3E) analyses, and multi-objective optimization of a CO_2/NH_3 cascade refrigeration system. *Appl. Therm. Eng.* **2014**, *65*, 42–50. [CrossRef]
28. Jain, V.; Sachdeva, G.; Kachhwaha, S.S. Thermo-economic and environmental analyses based multi-objective optimization of vapor compression–absorption cascaded refrigeration system using NSGA-II technique. *Energy Convers. Manag.* **2016**, *113*, 230–242. [CrossRef]
29. Wang, Y.; Wang, C.; Feng, X. Optimal match between heat source and absorption refrigeration. *Comput. Chem. Eng.* **2017**, *102*, 268–277.
30. Chen, J.J.J. Letter to the editors: Comments on improvement on a replacement for the logarithmic mean. *Chem. Eng. Sci.* **1987**, *42*, 2488–2489. [CrossRef]

Article

Estimation of Ice Cream Mixture Viscosity during Batch Crystallization in a Scraped Surface Heat Exchanger

Alejandro De la Cruz Martínez [1], Rosa E. Delgado Portales [1], Jaime D. Pérez Martínez [1], José E. González Ramírez [2], Alan D. Villalobos Lara [3], Anahí J. Borras Enríquez [4] and Mario Moscosa Santillán [1,*]

1   Facultad de Ciencias Químicas, Universidad Autónoma de San Luis Potosí, Av. Dr. Manuel Nava #6, Zona Universitaria, San Luis Potosí C.P. 78210, S.L.P., Mexico; alex_ixoye23@hotmail.com (A.D.l.C.M.); rdelgado@uaslp.mx (R.E.D.P.); jdavidperez@hotmail.com (J.D.P.M.)
2   AlphaGary Corp., División Compuestos y AF/Plastificantes, Autopista Altamira km. 4.5, Puerto Industrial Altamira, Tamaulipas C.P. 89608, Mexico; jose.gonzalez.ramirez@alphagary.com
3   Departamento de Ingeniería Química, División de Ciencias Naturales y Exactas, Universidad de Guanajuato, Noria Alta s/n, Guanajuato C.P. 36050, Gto., Mexico; alanred10@hotmail.com
4   Centro de Investigación y Asistencia en Tecnología y Diseño del Estado de Jalisco A.C., Camino al Arenero 1227, El Bajío, Zapopan 45019, Jalisco, Mexico; anajo_borras@hotmail.com
*   Correspondence: mario.moscosa@uaslp.mx; Tel.: +52-444-826-2440 (ext. 6596)

Received: 31 December 2019; Accepted: 28 January 2020; Published: 3 February 2020

Abstract: Ice cream viscosity is one of the properties that most changes during crystallization in scraped surface heat exchangers (SSHE), and its online measurement is not easy. Its estimation is necessary through variables that are easy to measure. The temperature and power of the stirring motor of the SSHE turn out to be this type of variable and are closely related to the viscosity. Therefore, a mathematical model based on these variables proved to be feasible. The development of this mathematical relationship involved the rheological study of the ice cream base, as well as the application of a method for its in situ melting in the rheometer as a function of the temperature, and the application of a mathematical model correlating the SSHE stirring power and the ice cream viscosity. The result was a coupled model based on both the temperature and stirring power of the SSHE, which allowed for online viscosity estimation with errors below 10% for crystallized systems with a 30% ice fraction at the exit of the SSHE. The model obtained is a first step in the search for control strategies for crystallization in SSHE.

Keywords: viscosity; crystallization; ice-cream; modelling; scraped surface heat exchanger

---

1. Introduction

In the ice cream crystallization process, viscosity is the most evolving property. This is due to the liquid transforming into a semi-solid as a result of the occurrence of ice crystals. Clarke [1] described that ice cream mixture viscosity increases for two reasons: the liquid mixture temperature decreases and the volume fraction of solid particles (ice crystals) increase. Thus, ice cream viscosity is determined by the non-frozen phase viscosity and the ice fraction generated during its freezing. For an ice cream liquid mixture, apparent viscosity values have been reported between 0.1 and 0.8 Pa·s, at a shear rate of 115 s^{-1} [2,3]. For already crystallized ice cream, there are not many references about viscosity values. However, most authors agree that their rheological parameters, such as storage and loss modules (G′ and G″), increase exponentially as temperature decreases and ice fraction increases. This effect translates into an exponential viscosity increase [4–6].

To understand the evolution of ice cream mixture viscosity, its complex composition must be considered. In fact, ice cream is a mixture of air, water, milk fat, non-fat milk solids, sweeteners, stabilizers, emulsifiers, and flavorings. This mixture is partially frozen through water crystallization, which is the critical stage during its elaboration [1,7,8]. Thus, from a physicochemical point of view, ice cream is a complex colloidal system, considered at the same time as an emulsion, a dispersion, and a foam. These structures play an important role to determine the product quality and shelf life [9–12].

It has been determined that the rheological behavior of an ice cream corresponds to a shear thinning non-Newtonian fluid [2,13]. Furthermore, this fluid acquires a viscoelastic behavior as the temperature decreases and the concentration of the crystals increases. This change in rheological behavior is due to the complex microstructure mixture that evolves significantly with temperature changes, which is the main crystallization promoter during processing and storage recrystallization [6,9,14].

There are some studies that propose mathematical models to describe the viscosity evolution of ice cream mixtures during crystallization [13,15]. All these models are based on the viscosity estimation equation for suspensions with rigid spherical particles proposed by Einstein, as described below:

$$\eta_{app} = \eta_o(1 + 2.5\phi) \tag{1}$$

where η_{app} is the apparent ice cream viscosity, η_o is the non-frozen liquid mixture viscosity, and ϕ is the solids volumetric fraction. This model has been the basis to determine the viscosity of different suspensions at low solid concentrations [16]. Using this model as a basis and seeking to extend its range of use, Thomas proposed a modification to the Einstein equation [17]. The modified equation is expressed as follows:

$$\eta_{app} = \eta_o\left[1 + 2.5\phi + 10.05\phi^2 + 0.00273\exp(16.6\phi)\right] \tag{2}$$

This model is valid for a solid fraction up to 0.625 and for a particle size range between 0.1 and 435 μm. This equation has been widely used by several authors to study ice suspensions [13,15,18] Equation (2) has been even used to estimate viscosity evolution in ice cream production, but it has been found to be valid only for solid concentrations below 10% [16,17]. For solid concentrations higher than 10%, the authors propose a parameters readjustment. Arellano et al. (2013) measured the viscosity of a sorbet at the SSHE exit using the capillary viscometer technique. The ice fraction was estimated knowing the sorbet temperature and by considering the equilibrium liquidus curve. Hence, once experimentally determined the sorbet's rheological parameters, using the Thomas model, the system's apparent viscosity was estimated with an error percentage of ±20%. Benkhelifa et al. [19] measured the viscosity of sorbets and ice cream by using a rotational viscometer inserted in a batch SSHE and by applying a Couette analogy in this non-conventional geometry.

There are also studies that relate ice cream viscosity with the stirrer energy consumption inside the scraped surface heat exchangers (SSHE). These models allow to estimate the power consumption of scraping blades as a function of ice cream viscosity during crystallization. The developed models consider that mixture viscosity increases as the crystal content increases. So, a direct proportional relationship between the energy consumption of the SSHE stirrer and apparent viscosity is proposed [1]. This increase in engine power leads to a friction increase between the scrapers and the internal wall of the SSHE, requiring a complementary energy input to the system to keep a constant agitation rate during the process. This extra energy is dissipated in the ice cream and later it is also removed by the cooling system [13,17].

One of the first models that relates an SSHE system viscosity with the energy consumed was developed by Skelland and Leung [20]. The model relates the power number (P_o) with the rotational Reynolds number $Re_R(= \rho N_R d_t^2 / \eta)$ and the number of scraping vanes (F). Thus, for the freezing of a glycerol-water solution, the proposed equation is:

$$P_o = 77,500\,Re_R^{-1.27}F^{0.59} \tag{3}$$

Trommelen and Beek [21] established that in the SSHE the power number was not solely a function of the rotational Reynolds number (Re_R), since their experimental results showed that for the same Re_R, the P_o was increased with a decrease of apparent viscosity. Therefore, they proposed the empirical equation:

$$P = \frac{251(N_R d_t)^{1.79} \eta^{0.66} F^{0.68} L_B}{(d_t - d_s)^{0.31}} \tag{4}$$

where P is the power consumed by agitation, N_R is the rate of rotation of the scraping blades, η is the fluid apparent viscosity, L_B is the length of the scraping blades, while d_t and d_s are the diameter of the tank and the stator of the SSHE, respectively. This model showed a proper agreement with experimental data ($R^2 = 0.992$).

In another study, Trommelen and Boerema [22], proposed a modification to the Skelland and Leung model. The proposed model can be expressed as:

$$\frac{P_o}{L_B} = K\, Re_R^{-a} F^{0.59} \tag{5}$$

where the rotational Reynolds number is a function of the viscosity ($Re_R = \rho N_R d_t^2 / \eta$), K and a are model fitted constants. This model was retaken by Qin et al. (2006 and 2007) to propose a predictive model of apparent viscosity during the crystallization of sucrose aqueous solutions. The proposed methodology consists in estimating the apparent viscosity, during crystallization, using the Thomas predictive model (Equation (2)). Complementarily, the stirrer torque in the SSHE was measured and the consumed power was calculated. Subsequently, the power number was estimated $\left(P_o = P/\rho N_R^3 d_t^5\right)$. Finally, the obtained values were used to find the adjusted parameters of the Trommelen and Boerema model (Equation (5)). Predictive curves were obtained and compared with experimental data, obtaining a proper fit. It is important to note that these tests were performed with sucrose solutions in water, with a less than 15% ice crystals volume fraction.

In conclusion, the models found in the literature, for the prediction of the apparent viscosity of mixtures during crystallization by freezing in SSHE's, have been developed mainly for sugar solutions and sorbets. The current challenge is to develop new models for complex mixtures, such as ice cream, considering a concentration of solids (ice crystals) up to 50% in the outlet stream. Therefore, the objective of this work focuses on the development of a viscosity predictive model for an ice cream mixture during its batch crystallization in a SSHE. This model could serve to build an online virtual instrument for viscosity estimation as a starting point to develop new control strategies for ice cream process production.

2. Materials and Methods

2.1. Ice Cream Mixture and SSHE Description

The ice cream was made using a branded commercial mixture from Golden Abarca®(La Perseverancia, Mexico). The mixture is composed of 15.4% carbohydrates (10.6% added saccharose and 4.8% lactose from milk), 5.2% fat, 2.7% proteins, 76.1% water and 0.6% ash. The SSHE consists of a jacketed cylindrical tank with a capacity of 500 mL. This tank is coupled with a primary cooling system using R404-A refrigerant and a secondary cooling bath using ethylene glycol, this system allows to reach a bath temperature of −30 ± 0.5 °C. The cooled ethylene glycol circulates inside the SSHE jacket as shown in Figure 1. The scraping blades are coupled to a Vexta®motor, model BLFM5120-GFS, equipped with a rotation rate control. Stirring rate can be controlled in the range of 0–800 rpm. In addition, a Hioki®power meter model 3286-20, connected to the motor power supply line, is used to record energy consumption during the process. The equipment is also equipped with a set of T-type thermocouples located at critical points of the cooling system, as well as inside the

SSHE, for temperature measurement. These thermocouples are connected to a 34970A Agilent®data acquisition system.

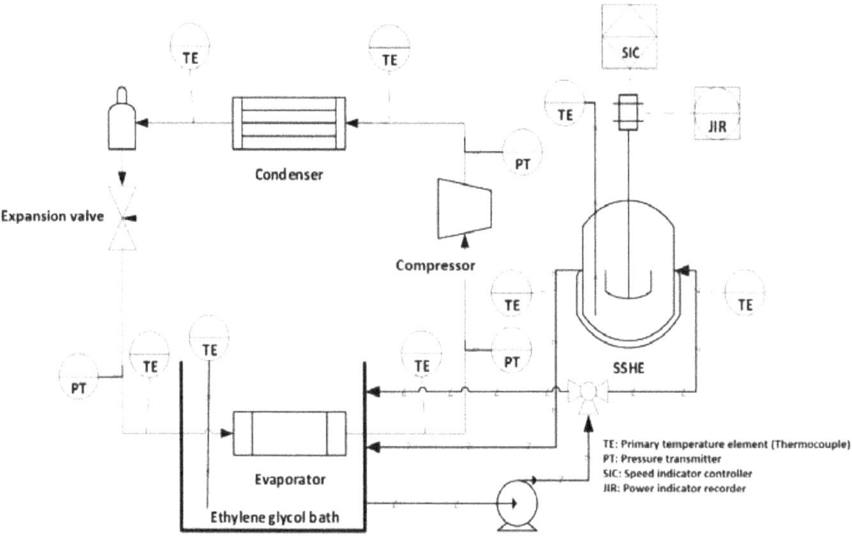

Figure 1. Scheme of the SSHE coupled to a cooling system.

2.2. Ice Cream Production

For each experiment, 300 mL of ice cream mixture were frozen using three stirring rates (73.78, 129.18 and 184.58 s^{-1}). The temperature of the cooling bath was kept at −30 °C, which allowed to control the mixture temperature in a range of 20 to −5 °C. Temperature was measured and recorded for each experiment as a function of time. Similarly, the energy consumption (power) of the stirring motor was recorded for each test. The ice fraction was estimated from the temperature of the mixture during the crystallization stage. For this purpose, the ice cream crystallization temperature curve at different concentrations, obtained by means of differential scanning calorimetry (DSC), using a TA Instrument®calorimeter model Q2000 coupled to a cooling system model RCS90, was used.

2.3. Rotational and Oscillatory Rheology for Ice Cream Mixtures during Their Transformation

Experiments were carried out to measure the rheological properties of the mixture and the frozen ice cream, according to the methodology for the rheological characterization of ice cream proposed by Anton Paar® [23]. For this, a Physical MCR 302 rheometer from Anton Paar®was used with a stainless-steel parallel plate geometry model PP25-P/TG with TruGap^{TM} (25 mm diameter). The geometry is coated with a ceramic chrome oxide (Cr$_2$O$_3$) layer. In order to prevent slip, a profiled geometry was used. The temperature was controlled with a Peltier PTD 150 unit. A gap of 2 mm and a deformation constant of 0.02% was programmed for all experimental runs. At the beginning of each run the temperature was adjusted to 20 °C, maintaining control by means of the Peltier unit. Subsequently, the evolution of the apparent viscosity of the mixture during crystallization was determined. For this purpose, crystallization runs were carried out in the SSHE, which were interrupted at different temperatures (0, −1, −2, −3, −4, −5 °C) to take a sample for each temperature and determine the apparent viscosity. All experiments were carried out twice to verify repeatability. It should be mentioned that necessary measures were taken to avoid the melting of the samples in their passage from the SSHE to the rheometer, placing the SSHE as close as possible to the rheometer, as well as keeping the rheometer temperature at the sample temperature.

It should be noted that, for the first three temperatures (0, −1, −2 °C), the mixture was obtained as a liquid. So, the rheological tests were carried out in rotational mode. To do this, for each temperature a cut rate sweep was made over the sample from 1 to 500 s^{-1} ($\dot{\gamma}$). Then, using experimental data, the evolution of apparent viscosity was determined (η_{app}). Typically, for each rheological analysis 50 measurement points were obtained with an average interval of 6 s between them.

For temperatures of −3, −4, and −5 °C, the mixture showed a certain degree of freezing. So, an oscillating measurement mode was chosen. For this purpose, at each temperature sampled, an angular frequency sweep from 1 to 500 s^{-1} was performed. The rheological results obtained were complex viscosity (η^*) as well as loss G'' and storage $G\prime$ modules. In this case, the experimental data obtained represent a total of 100 measuring points with an average sampling time of 6 s. These data were then used to estimate dynamic viscosity (η').

2.4. Oscillatory-Rotational Mode Tests for Viscosity vs. Temperature Measurement

To determine the ice cream viscosity evolution during crystallization, three thermo-rheological tests from the ice cream melt were carried out. A combination of oscillatory and rotational mode was used in the rheometer, using a temperature sweep of −10 to 10 °C [4]. Each run corresponds to a frequency (Hz) and a shear rate (s^{-1}) equivalent to the stirring rate used in the SSHE (73.78, 129.18, and 184.58 s^{-1}). In these experiments, samples of ice cream frozen at −18 °C, with more than 24 h of storage, were used. The rheometer was conditioned to a temperature of −10 °C at the beginning of each run and the sample was placed.

The test started in oscillatory mode at constant frequency equivalent to the respective stirring rate (73.78, 129.18 and 184.58 s^{-1}). The oscillatory mode was used from −10 °C to −2.5 °C, with a heating rate of 0.25 °C/min, with a sampling time of 30 s taking 60 measuring points. Subsequently, the rheometer operation was switched to rotational mode at a constant shear rate equivalent to each stirring rate (73.78, 129.18, 184.58 s^{-1}). Thus, heating was continued from −2.4 °C to 10 °C with a heating rate of 1.2 °C/min, and a sampling time of 12 s taking 50 measuring points. As a result, complex viscosity (η^*), as well as storage and loss modules (G' y G'') were obtained for the oscillatory mode and dynamic viscosity was calculated from these parameters (η'). Alternatively, from rotational mode operation, the apparent viscosity was calculated (η_{app}).

3. Results and Discussion

3.1. Temperature, Stirring Power and Ice Fraction Monitoring

Figure 2 shows temperature evolution for the mixture during ice cream preparation. It is noted that the room temperature at the beginning of the process is approximately 20 °C. From this temperature it begins a gradual decrease, it is called "cooling stage". At this stage, the temperature decreases rapidly, reaching values close to −5 °C, corresponding to a mixture temperature below its initial crystallization temperature (T_c). Subsequently, a slight increase in temperature is observed until its T_c. Finally, a continuous decrease is observed, but with a slower cooling rate, and this stage corresponds to the crystallization process of ice fraction.

As is known, the crystallization is an exothermic process that provokes a sharp temperature increase until reaching the T_c, in this study the increment was −2.8 °C, as shown in Figure 2. The results of T_c at different solid concentrations were obtained by DSC analysis (Figure 3) and compared with the experimental results obtained in the SSHE (Figure 4). As shown in Figure 4, the T_c depends on the formulation of the mixture, mainly on the concentration of solids, and it is from this temperature that the formation of the ice fraction begins [16]. The slower temperature decrease observed after the initial crystallization point can be explained by the exothermic contribution of the phase change, the energy generated by the scraping that increases as solids concentration increases, and also by the increase in viscosity which decreases the heat dissipation rate [24,25].

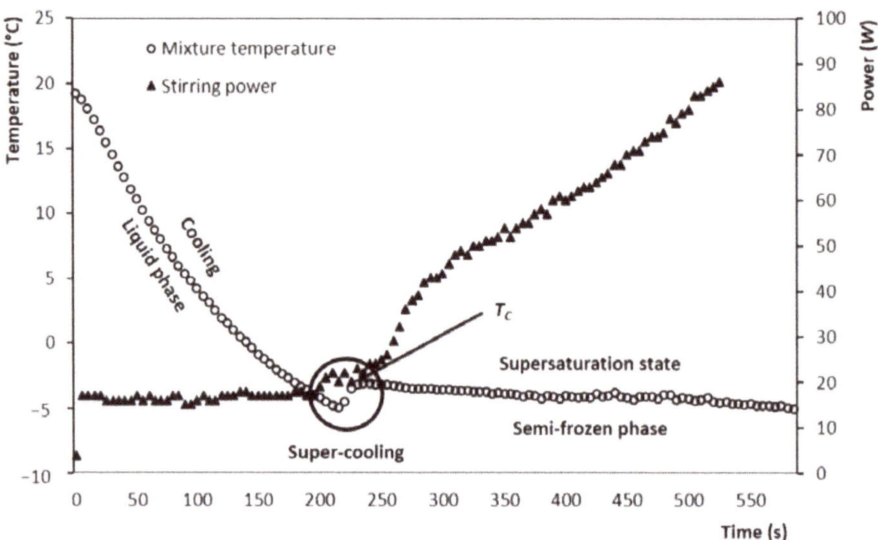

Figure 2. Evolution of the mixture temperature (▲) during crystallization in the SSHE and evolution of the power consumed by stirring motor (○), test performed at 200 rpm (3.33 s^{-1}).

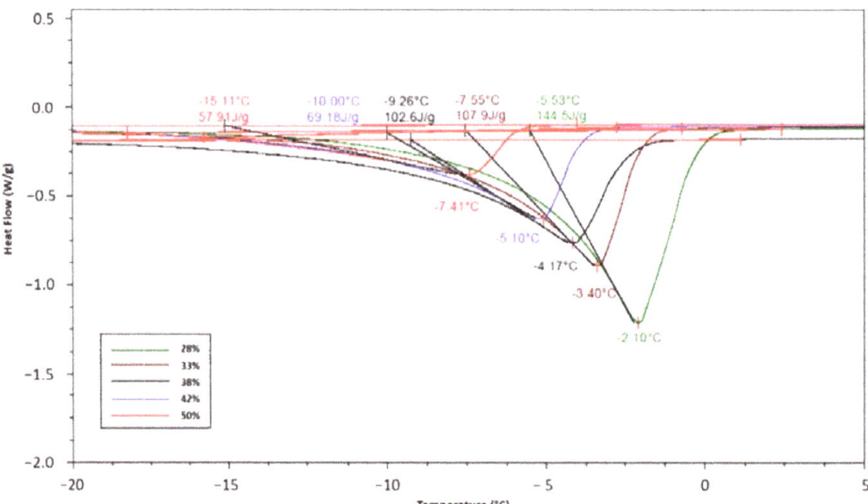

Figure 3. T_C values determined by DSC analysis at different solid concentrations in the ice cream mixture.

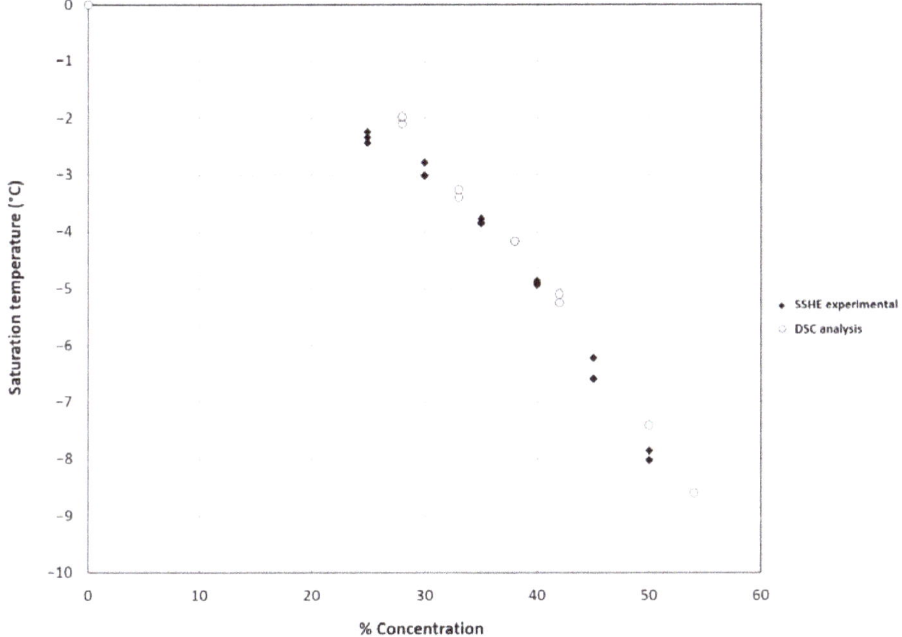

Figure 4. Comparison of experimental T_c values obtained in the SSHE and DSC analysis results of the ice cream mixture as a function of solid concentration (%).

The evolution of the power consumed by the stirring motor of the SSHE is also shown in Figure 2. It is observed that after the T_c occurs, the power consumed by the stirrer rotation increases significantly, as the temperature drops, and the ice fraction increases (Figure 5). Consequently, the highest value of power consumption (~90 W) was found at the end of the crystallization (−5 °C, 600 s). In the first stage, the stirrer had a continuous consumption of 15 W, to maintain a constant stirring rate of 3.33 s^{-1} for the liquid mixture. During the crystallization process, this energy consumption gradually increased as the temperature of the semi-frozen phase decreased. The end of the process was established at −5 °C and the fraction of ice was estimated as a function of temperature, according to the methodology proposed by Cerecero (2003). This methodology is based on the solute mass balance considering that there are only water crystals. The mathematical relationship considers the initial mass fraction of solute and the residual mass fraction related to temperature by the liquidus curve (Figure 4). The increase in the ice fraction, for this same process, can be seen in Figure 3. It is observed that, like the power, the increase in the ice fraction starts at temperatures below the Tc. This increase in the ice fraction (Figure 5) corresponds to the increase in the power consumed for the agitation of the semi-frozen phase (Figure 2). At the end of the process, an approximate ice fraction of 30% was obtained. To keep the stirring rate constant with this ice concentration, the energy consumption increased to values close to 90 W. As can be seen, the increase in power consumption is six times greater than the power required for stirring the liquid mixture. It should be noted that the heat generated by these changes is dissipated in the ice cream and finally removed by the cooling system [1,2,15,26].

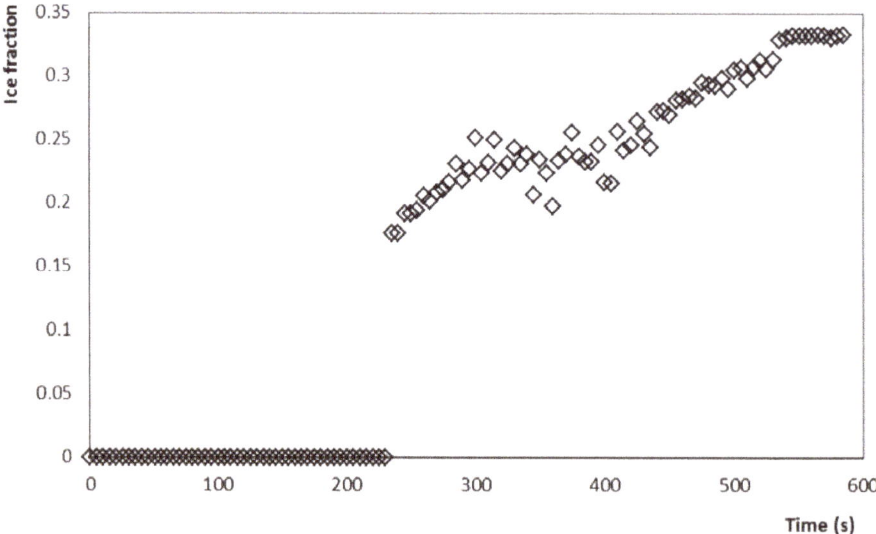

Figure 5. Ice fraction increase during crystallization of the mixture up to −5 °C at a stirring rate of 200 rpm (3.33 s^{-1}).

This dynamic crystallization process is similar to that described by several authors, who have studied ice cream crystallization in SSHE [2,7,10,13,15,25,27–30]. The evolution of the temperature and therefore the ice fraction were similar in all the tests carried out at different agitation rates.

The results show that the energy increase in the power consumed by the stirring motor of the SSHE is proportional to the increase in the fraction of crystals formed, and therefore must be related to the increase in the viscosity of the ice cream during its crystallization. So, it is possible to find the relationship that manages to estimate viscosity evolution based on the power consumed by a SSHE engine.

3.2. Rheological Results

3.2.1. Behavior of the Mixture in Liquid and Crystallized State

The ice cream mixture was kept in a liquid state when it was above −2.8 °C, and although its viscosity increased slightly with the decrease in temperature, this increase is not significant (0.16%). In addition, the results at temperatures of 0, −1, and −2 °C (Figure 6) show that the apparent viscosity of the mixture is dependent on the shear rate, this being a typical characteristic of a non-Newtonian fluid.

The liquid mixture near the freezing point behaves as a shear-thinning pseudoplastic fluid, which has already been reported by other authors [2,31]. This behavior is due to the so-called thixotropy, which is caused by the breakdown of interactions and associations (weak bonds) between the different components of the mixture. This thinning behavior is well described by the power law (Equation (6)). From the experimental results, the values of the consistency coefficient k and the flow behavior index n (Table 1) were obtained.

$$\eta_{app} = k \cdot \dot{\gamma}^{n-1} \tag{6}$$

Table 1. Values of k and n for the ice cream mixture at different temperatures.

T (°C)	k	n
0	6.7935	0.49605
−1	7.24805	0.4927
−2	7.69055	0.48755
−3	4.5118	0.4309
−4	23.008	0.2204
−4.5	43.4115	0.2464

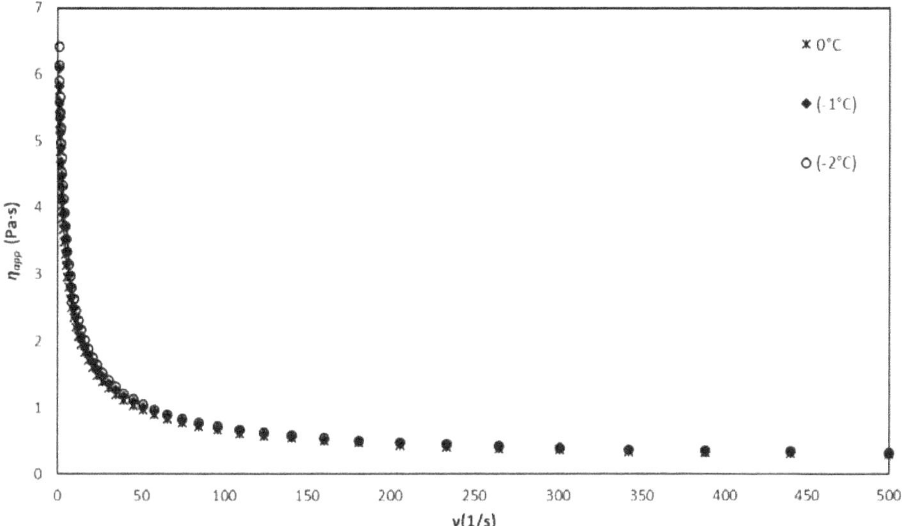

Figure 6. Viscosity of the liquid mixture vs. shear rate at temperatures of 0, −1, −2 °C.

When the mixture reaches below −2.8 °C, the appearance and growth of ice crystals is noticeable in the values obtained from dynamic viscosity measurements, for temperatures of −3, −4 and −4.5 °C (Figure 7).

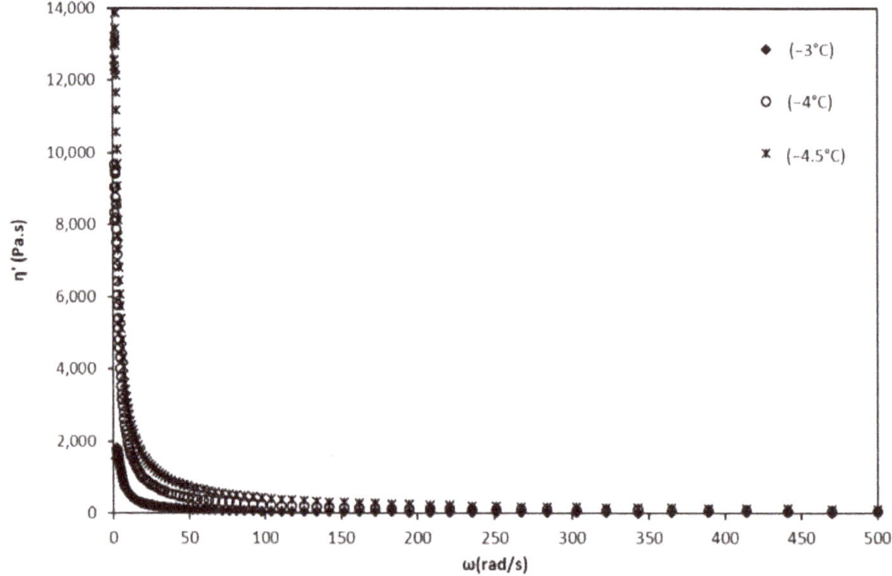

Figure 7. Dynamic viscosity vs. angular frequency at temperatures of −3, −4, and −4.5 °C.

The behavior of the mixture changed radically since it was transformed into a semi-solid system with a viscoelastic behavior [12]. This behavior, in the presence of crystals, prevents the apparent viscosity from being evaluated. Therefore, the complex viscosity was determined and from this the dynamic viscosity. In Figure 7 it can be seen that as the temperature decreases, the dynamic viscosity has increasing values. This dynamic viscosity was similarly adjusted to the power law to determine the values of the k and n indices, which are also shown in Table 1.

3.2.2. Ice Cream Melting for Evaluating the Dynamic and Apparent Viscosity

Ice cream melting tests in the rheometer showed a very similar behavior to that already described in other investigations [4,5]. Three zones are identified when evaluating G' and G'' modules, these zones can be seen in Figure 8 and are described below.

Zone 1: It is in low temperatures between −20 °C and −10 °C, where the microstructure of the ice crystals dominates. The storage module G', which describes the behavior of a solid body, has a greater decrease as the theoretical ice fraction decreases. While the loss module G'' which describes the viscous behavior is shown to be constant in this temperature range, this can be correlated with the rigidity and the formation of the ice cream ball when spooning it.

Zone 2: Includes the temperature range of −10 °C to −2.8 °C, where the theoretical ice fraction decreases significantly. Here, there is a slope of both modules that has a steep inclination, which refers to the melting rate of the ice crystals. Therefore, it can be correlated with the sensory attribute of cold when consuming ice cream.

Zone 3: It goes from the temperature range greater than T_C where both modules, G' and G'', enter their lowest levels and remain constant. There are no ice crystals here, so the foamy structure formed by the air and the fat become the dominant phase. The level at which the modules are can be related to the creaminess attribute.

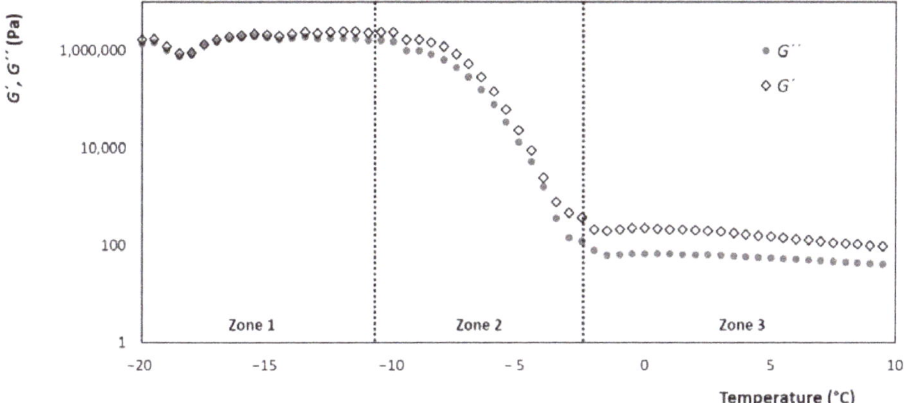

Figure 8. Evolution of the storage and loss modules (G' and G'' respectively) of ice cream, with respect to the change in temperature, with a constant frequency and shear rate of 68.93 s^{-1}.

Dynamic crystallization, carried out in SSHE, includes temperatures ranging from 5 to -5 °C, placing this process in zones 2 and 3 described above. For this region of temperatures, the modules are superimposed, as long as it is below the T_C, making the predominant behavior to be viscoelastic. This leads to the use of dynamic viscosity as the most appropriate parameter for temperatures where there is an ice fraction ($T < T_C$). However, when the $T > T_C$ the mixture is completely at a liquid state and its apparent viscosity can be estimated, from rotational tests, applying the power law.

Thus, from the thermo-rheological melting tests, it was possible to obtain the evolution of viscosity with respect to temperature. The result is shown in Figure 9, where there are two defined zones. One corresponding to the frozen part described by the ice cream dynamic viscosity η', in the range of -10 to -2.8 °C. Meanwhile, for the liquid part, it is described with the apparent viscosity η_{app} and corresponds to temperatures greater than -2.8 °C. The zone of union between both behaviors is considered a critical zone for the development of the model. This zone has a curvature that is dependent on the heating rate at which the rheometer was adjusted, during the fusion test. Therefore, since melting and crystallization are opposite phenomena, it is expected that this area will not be fully adjusted. In the crystallization process, the appearance of the ice fraction is instantaneous. On the other hand, in the fusion, the melting of the crystals is a gradual change that ends until reaching the T_C. Therefore, for this area, the viscosity data obtained during the rheological tests were used to determine k and n, which were explained in the previous section.

In Figure 7, three slopes are observed during the temperature increase. These denote the three areas where the ice cream structure is evolving in different ways, and primarily due to the disappearance of ice crystals. The viscosity decreases dramatically as the temperature increases in the range of -10 to -2 °C, which is the area where the fraction of crystals melts. From the range of -10 to -5 °C, the viscosity gradually decreases. Here, the ice crystals melt slowly. However, from -5 °C and until T_C the viscosity decreases dramatically, because the rate at which the crystals melt increases sharply [29,32]. Around the T_C and towards higher temperatures, the viscosity value stabilizes, as the ice fraction has completely disappeared.

The results obtained corroborate the relationship between the ice fraction, the power consumed by the SSHE engine, and the viscosity of the system. Each of these, increases as the mixture's temperature decreases during crystallization. Therefore, it is possible to develop a model that allows for the online estimation of viscosity based on the power consumed by the SSHE, which is an easy measurement parameter.

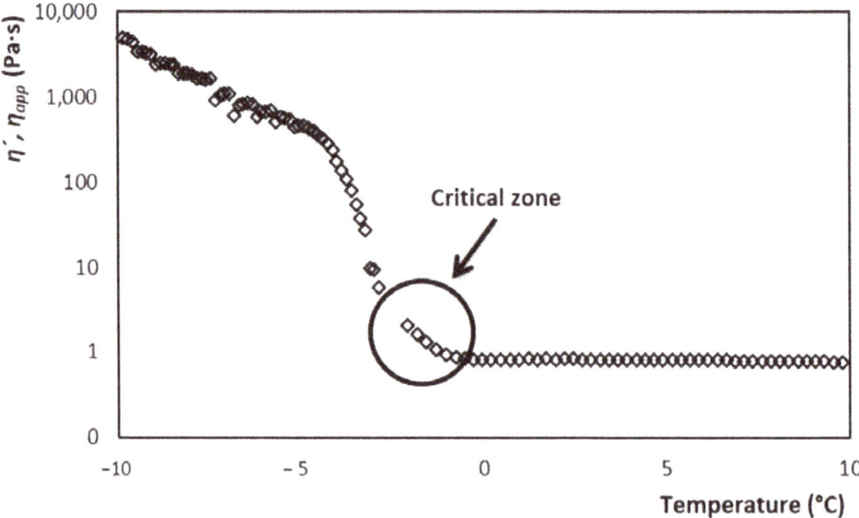

Figure 9. Evolution of dynamic viscosity (−10 to −3 °C) and apparent viscosity (−2.5 to 10 °C) of ice cream, with respect to change temperature, with a constant frequency and shear rate from 73.78 s^{-1}.

4. Model Development for Viscosity

4.1. Initial Considerations

A mathematical model was developed for viscosity estimation, contemplating two stages during the crystallization process: the first is a cooling stage (starting in the initial temperature range to the initial point of crystallization) while the second stage contemplates the appearance and crystals growth (from the initial point of crystallization to −5 °C). The two stages can be seen in Figure 10; the cooling stage is represented by an apparent viscosity (η_{app}), and the crystallization stage by dynamic viscosity (η'). The evolution of both viscosities was obtained from melting tests of an ice cream sample at a constant shear rate, using a rheometer, according to the methodology proposed by Wildmoser et al. (2004). For the critical zone shown in Figure 9, the values obtained from the fusion were replaced, with values determined at the end of interrupted crystallizations at different temperatures (0, −1, −2, −3, −4, and −4.5 °C). These values adjusted very well to the two stages described by the fusion. But the most significant event is that the critical zone was filled with real crystallization values, which is the path that truly follows the ice cream viscosity during its processing in the SSHE.

The evolution of viscosity during crystallization at a stirring rate of 3.33 rev/s in the SSHE is represented in Figure 10. This is the behavior that was estimated from the mathematical model. As you can see, the behavior consists of two stages, the first is a cooling stage and the second is the crystallization stage. For the cooling stage, the apparent viscosity values obtained from the rheometer through a rotational mode fusion were used. For the crystallization stage, the dynamic viscosity values obtained from rheometer tests which consisted of a fusion in oscillatory mode were used. Point viscosity values were obtained in the transition zone between both stages; for this, crystallizations were performed which were stopped at different temperatures and from which a sample was taken and brought to the rheometer for viscosity determination (black dots in Figure 10).

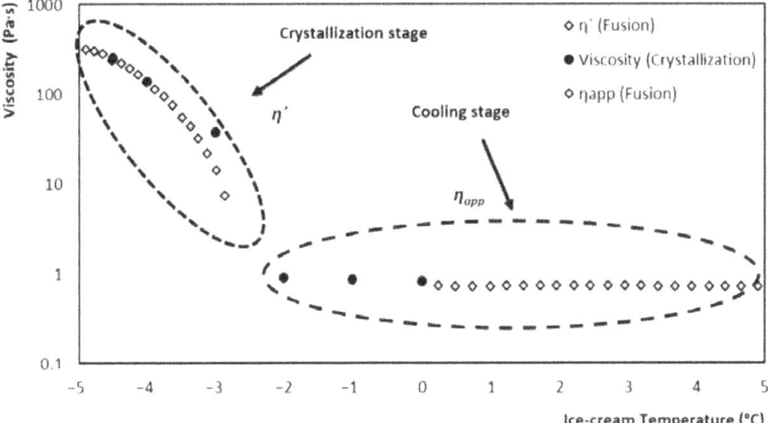

Figure 10. Evolution of ice cream apparent and dynamic viscosity: at a stirring rate of 3.33 rev/s. Crystallization stage (−2.8 °C to −5 °C), cooling stage (−2.8 °C to 5 °C).

A scheme of the model shaping strategy is shown in Figure 11, where t represents the time at which crystallization took place. The study was delimited at a temperature range (T) of 5 to −5 °C, highlighting that the T_C of the mixture was found at −2.8 °C. In this range, the temperature and power measurement (P) was made. These experimental values are used as model variables. For the cooling stage, the viscosity is estimated from the temperature using the power law model, resulting in an apparent viscosity (η_{app}). Once the temperature reaches the T_C, the power law model is replaced by a function based on the power consumed by the motor; which allows to estimate the dynamic viscosity (η') during the crystallization stage. The model was validated with measurements of apparent viscosity as a function of the shear rate ($\dot{\gamma}$) for the first stage, and dynamic viscosity measurements as a function of the loss module G'' for the second stage. Thus, the global model for viscosity prediction is shaped by coupling these two functions.

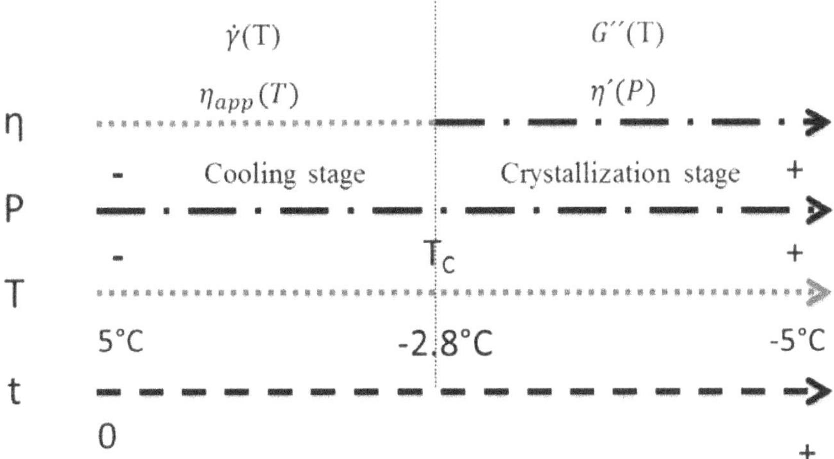

Figure 11. Representation of the model's two stages of action. T_C is the mixture's crystallization temperature, P is power consumed by the stirring motor, η is the viscosity (η_{app} is apparent viscosity, η' is dynamic viscosity), t is time, $\dot{\gamma}$ is velocity of cut and G'' is the loss module.

4.2. Calculation of Average SHEAR Rate in the SSHE

The determination of the average shear rate, within the SSHE, is a function of the scraping blades agitation rate. There are different models that focus on the calculation of an average shear rate for this type of equipment [33], Yataghene et al. [34] make a compendium of these models. Another methodology to estimate an average shear rate in the SSHE is using the Couette analogy described by Steffe [35], and used by Benkhelifa et al. (2008). The proposed by Mabit et al. [36] is one of the most employed. This model consists in estimating the shear rate in two important areas within the SSHE. The first zone is on the rotor (Equation (7)). The second corresponds to the hollow space between the cylinder and the rotor (Equation (8)). From the two cut rates, an overall shear rate of the SSHE is calculated as an average (Equation (9)). Table 2 shows the calculated values for each of the rates used in this study. In addition, the values of the power consumed at the beginning and end of each run are presented, as well as the final temperature and the estimated ice fraction.

$$\dot{\gamma}_{rotor} = \frac{2\pi R_r^{2/n}}{R_s^{2/n} - R_r^{2/n}} \left[1 - \frac{2-n}{n} \left(\frac{R_s}{R_r} \right)^{2/n} \right] N_R \tag{7}$$

$$\dot{\gamma}_{Bowls} = \frac{4\pi}{n} \left[\frac{R_r^{2/n}}{R_s^{2/n} - R_r^{2/n}} \right] N_R \tag{8}$$

$$\tilde{\dot{\gamma}} = \frac{\dot{\gamma}_{Bowls} + \dot{\gamma}_{rotor}}{2} \tag{9}$$

Table 2. Estimated shear rate values for different SSHE agitation rates, power consumed at each rate and estimated ice fraction.

Velocity(s^{-1})	$\tilde{\dot{\gamma}}$ (s^{-1})	P_{star} (W)	P_{end} (W)	T_{out} (°C)	ϕ
3.33	73.78	14	61	−4.5	0.27
5.83	129.18	27	143	−5	0.3
8.33	184.54	42	169	−5	0.3

In both equations, R_r and R_s are the ratio of the rotor and the stator (internal radius of the cylinder), respectively. N_R is the rotational rate in s^{-1} and n was taken as an average of the different values obtained for this parameter in liquid state, which was 0.492, according to the values presented in Table 1.

4.3. Development of Model Equations

4.3.1. Estimation of Dynamic Viscosity from Power (Crystallization Stage)

Authors such as Skelland and Leung (1962), Trommelen and Beek (1971) and Qin *et al.* (2006), have estimated the energy consumption (P) in the SSHE by means of the relationship between variables such as, the length of the scraping blades (L_B), the number of scraping blades (F), and the number of Reynolds (Re). This relationship has been expressed in Equation (5), where P_o is the power number, while K and a are adjustment constants. For this study, $F = 4$ and $L_B = 0.126$ m. The P_o is given by the following equation:

$$P_o = P / \rho N_R^3 d_t^5 \tag{10}$$

The Reynolds number of the mixture within the SSHE was obtained by the equation:

$$Re = \rho N_R d_t^2 / \eta \tag{11}$$

The model was constructed by combining Equations (5) and (11). Thus, the apparent viscosity is a function of the power:

$$\eta\prime = \rho N_R d_t^2 \sqrt[a]{\frac{P_o}{KF^{0.59}L_B}} \tag{12}$$

Equation (12) represents the first part of the model and corresponds to the relationship of viscosity with the stirring motor increase in energy consumption. A maximum value of $\rho = 650$ kg/m^3 was considered for the ice cream density, which corresponds to the largest fraction of ice generated ($\phi = 0.3$). This ice cream density was calculated using the mixture density and the experimental overrun determined from production runs [2]. The internal diameter of the SSHE was $d_t = 0.073$ m. N_R as already mentioned is the rotational rate of the scraping blades (3.33, 5.83, 8.33 s^{-1}).

The values of the model adjustment parameters, represented in Equation (12), K and a, were determined from the experimental data using a nonlinear least squares regression method, according to the equation:

$$SSE = \frac{1}{M} \sum_{1}^{M} \left(\eta_{app.medida} - \eta_{app.estimada} \right)^2 \tag{13}$$

where M is the number of experimental points, $\eta_{app.measure}$ is the apparent viscosity and $\eta_{app.estimated}$ is the predicted value for the ice cream mixture apparent viscosity. Values of 16.73 and -1.38 were obtained for K and a, respectively.

The relative error between the experimental viscosity and the estimated viscosity was determined for limit values of $\pm10\%$ and is reported in Figure 12. It is observed that the dispersion of the results is within the delimited region. So, it can be stated that in any case, the estimation error is less than 10%. This confirms the coherence between the data estimated from the stirring power and the experimental viscosity data obtained in the rheometer. These results represent a fair estimate for the crystallization of a complex mixture such as ice cream base. Therefore, it can be said that the power-based model was accurate for the crystallization zone.

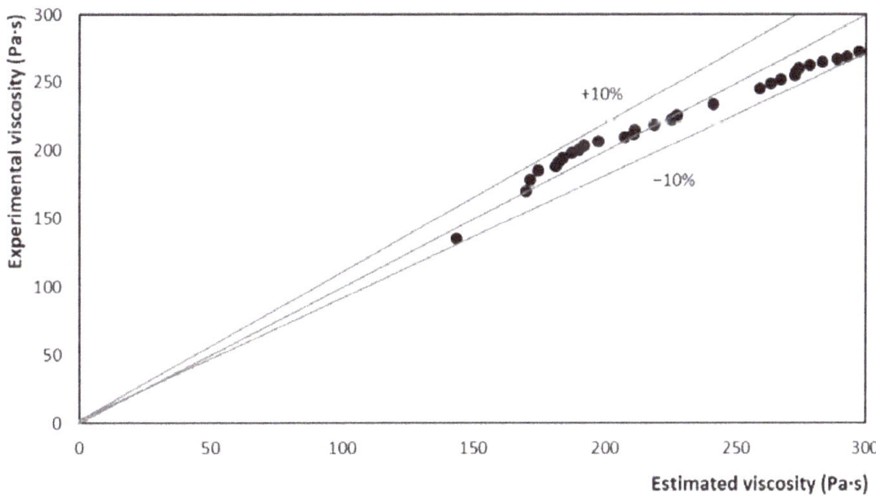

Figure 12. Comparison between experimental and estimated values for apparent viscosity, using the model as a function of the stirring motor power at 3.33 rev/s.

4.3.2. Estimation of Apparent Viscosity from Temperature (Cooling Stage)

At temperatures above the T_C, the mixture has no ice fraction, so the apparent viscosity of the mixture is only affected by the temperature and is determined based on the shear rate that is applied.

At this stage the behavior of the mixture is well described by the power law (Equation (6)), where k and n are the variables that characterize the mixture and are affected by temperature. Equation (6) represents the second part of the model, for the cooling stage, where the change in viscosity is minor and is only due to a change in temperature.

To determine the effect of temperature on the variables of Equation (6), Arrhenius's law was used, obtaining Equations (14) and (15), where A_k, E_k, A_n and E_n are adjustment parameters that were determined from experimental data, using Scilab®6.0 software, with the nonlinear least squares regression method. These data were the values of k and n, obtained at three temperatures at which the mixture still remained in a liquid state (0, −1, −2 °C) and the adjustment was made by means of a linear regression.

$$k(T) = A_k \cdot \exp\left(\frac{E_k}{T(°K)}\right) \tag{14}$$

$$n(T) = A_n \cdot \exp\left(\frac{E_n}{T(°K)}\right) \tag{15}$$

The values obtained for A_k, E_k, A_n and E_n are 3.5754×10^{-4}, 4578.4, 4.4965, and −602.16, respectively. These values are in the same order of magnitude of the values reported for other fluids with slimming behavior, such as ice cream mixture [37].

4.3.3. Coupling of Power and Temperature Models

The complete model of viscosity evolution is represented by Equations (6) and (12), with their respective parameters adjusted by experimental data. These equations were coupled using a Heaviside function that uses the crystallization temperature T_C (−2.8 °C) as the transition parameter. Equations (16) and (17) represent the final model.

$$\eta(T,\, P) = [H(T\prime) \cdot - 1]\rho N_R d_t^2 \sqrt[a]{\frac{P_o}{KF^{0.59}L_B}} + k(T) \cdot \dot{\gamma}^{n(T)-1} \tag{16}$$

$$T' = T - T_C \tag{17}$$

In this way, the predictive model for the entire temperature range used during crystallization in the SSHE was formed, being constituted by the power law, depending on the temperature, and the modified Qin model, depending on the power consumed.

The validation of the model was carried out with tests performed at different agitation rates, which correspond to the average shear rates of the SSHE to which the mixture is subjected during crystallization. Figure 13a shows the graph that contrasts the values estimated by the model throughout the crystallization run and the actual viscosity measurements obtained in the rheometer by the melt run, for a velocity of 3.33 s^{-1}. The model accurately describes the viscosity behavior at both stages of the process (cooling and crystallization). Similar results were obtained for the three case studies where the agitation rate was varied (5.83 and 8.33 s^{-1}), Figure 13b,c.

In the three cases, the power-based model managed to adjust very well to the experimental values, this crystallization stage being the one of greatest interest in the model. Therefore, it can be said that the model managed to adequately estimate the evolution of viscosity regardless of the operating conditions of the process, thus demonstrating with it its efficiency for the online measurement of viscosity in discontinuous SSHE.

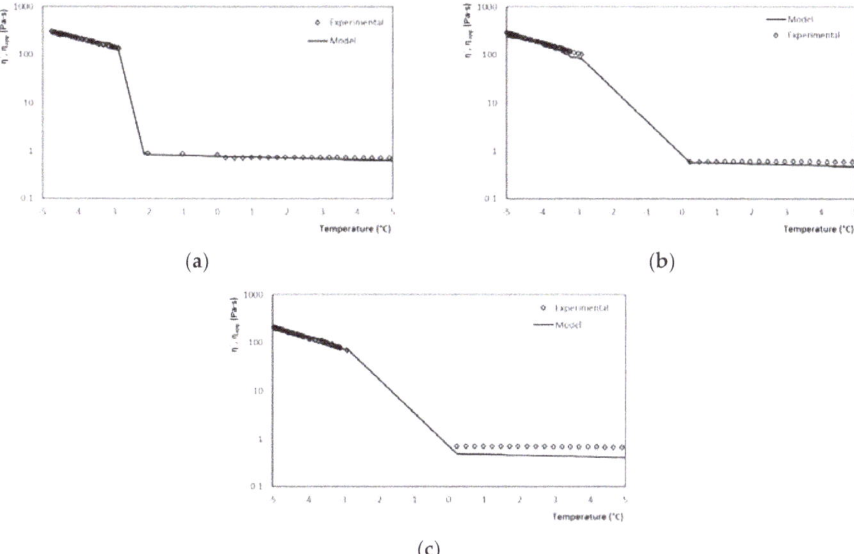

Figure 13. Comparison between experimental values and those estimated by the model for viscosity using the coupled model for rates of (**a**) 3.33, (**b**) 5.83 and (**c**) 8.33 s^{-1}.

5. Conclusions

This study found that the dynamic crystallization of ice cream in SSHE is the most important stage for the development of physical properties (ice fraction and viscosity), which determine the quality of the ice cream. With this work, a predictive model of the evolution of these properties was obtained, which was effective when tested at different stirring rates (<10% error). In this way, on-line estimation of the ice fraction and the ice cream viscosity was possible. Moreover, this was possible from the measurement of simple variables such as the stirring power consumption and the mixture temperature. Furthermore, the proposed model is valid for ice fractions up to 30%, which are typical values found in crystallization processes of this type of product, and which have not been reported in the literature. This represents a first step in establishing a control strategy for the ice cream crystallization process.

Author Contributions: Conceptualization, J.E.G.R., R.E.D.P. and M.M.S.; formal analysis, A.D.l.C.M., A.J.B.E., J.E.G.R. and M.M.S.; investigation, A.D.l.C.M., A.D.V.L. and A.J.B.E.; methodology, R.E.D.P., J.E.G.R., J.D.P.M. and M.M.S.; project administration J.E.G.R. and M.M.S.; resources R.E.D.P., J.D.P.M. and M.M.S.; supervision, R.E.D.P., J.E.G.R. and M.M.S.; visualization, A.D.l.C.M. and A.D.V.L.; writing—original draft, A.D.l.C.M.; writing-review & editing, R.E.D.P. and M.M.S. All authors have read and agreed to the published version of the manuscript.

Funding: This research was funded by FAI program from Universidad Autónoma de San Luis Potosí.

Acknowledgments: The authors acknowledge the use of the facility access in Food Pilot Plant and the Food Microbiology Laboratory, of the Universidad Autónoma de San Luis Potosí, where experiments were carried out for the development of the research. Especially to M.C. María del Refugio Pérez Barba and Ing. Alejandra Loredo Becerra.

Conflicts of Interest: The authors declare no conflict of interest.

References

1. Clarke, C. The physics of ice cream. *Phys. Educ.* **2003**, *38*, 248–253. [CrossRef]
2. Goff, H.D.; Hartel, R.W. *Ice Cream*; Springer: Boston, MA, USA, 2013.
3. Amador, J.; Hartel, R.; Rankin, S. The Effects of Fat Structures and Ice Cream Mix Viscosity on Physical and Sensory Properties of Ice Cream. *J. Food Sci.* **2017**, *82*, 1851–1860. [CrossRef]

4. Wildmoser, H.; Scheiwiller, J.; Windhab, E.J. Impact of disperse microstructure on rheology and quality aspects of ice cream. *LWT Food Sci. Technol.* **2004**, *37*, 881–891. [CrossRef]
5. Eisner, M.D.; Wildmoser, H.; Windhab, E.J. Air cell microstructuring in a high viscous ice cream matrix. *Colloids Surf. A Physicochem. Eng. Asp.* **2005**, *263*, 390–399. [CrossRef]
6. Tsevdou, M.; Gogou, E.; Dermesonluoglu, E.; Taoukis, P. Modelling the effect of storage temperature on the viscoelastic properties and quality of ice cream. *J. Food Eng.* **2015**, *148*, 35–42. [CrossRef]
7. Drewett, E.M.; Hartel, R.W. Ice crystallization in a scraped surface freezer. *J. Food Eng.* **2007**, *78*, 1060–1066. [CrossRef]
8. Parra, O.D.H.; Ndoye, F.T.; Benkhelifa, H.; Flick, D.; Alvarez, G. Effect of process parameters on ice crystals and air bubbles size distributions of sorbets in a scraped surface heat exchanger. *Int. J. Refrig.* **2018**, *92*, 225–234. [CrossRef]
9. Soukoulis, C.; Rontogianni, E.; Tzia, C. Contribution of thermal, rheological and physical measurements to the determination of sensorially perceived quality of ice cream containing bulk sweeteners. *J. Food Eng.* **2010**, *100*, 634–641. [CrossRef]
10. Cook, K.L.K.; Hartel, R.W. Mechanisms of ice crystallization in ice cream production. *Compr. Rev. Food Sci. Food Saf.* **2010**, *9*, 213–222. [CrossRef]
11. Varela, P.; Pintor, A.; Fiszman, S. How hydrocolloids affect the temporal oral perception of ice cream. *Food Hydrocoll.* **2014**, *36*, 220–228. [CrossRef]
12. Hartel, R.W.; Rankin, S.A.; Bradley, R.L. A 100-Year Review: Milestones in the development of frozen desserts. *J. Dairy Sci.* **2017**, *100*, 10014–10025. [CrossRef] [PubMed]
13. Arellano, M.; Flick, D.; Benkhelifa, H.; Alvarez, G. Rheological characterisation of sorbet using pipe rheometry during the freezing process. *J. Food Eng.* **2013**, *119*, 385–394. [CrossRef]
14. Guo, E.; Kazantsev, D.; Mo, J.; Bent, J.; Van Dalen, G.; Schuetz, P.; Rockett, P.; StJohn, D.; Lee, P.D. Revealing the microstructural stability of a three-phase soft solid (ice cream) by 4D synchrotron X-ray tomography. *J. Food Eng.* **2018**, *237*, 204–214. [CrossRef]
15. Qin, F.; Chen, X.D.; Ramachandra, S.; Free, K. Heat transfer and power consumption in a scraped-surface heat exchanger while freezing aqueous solutions. *Sep. Purif. Technol.* **2006**, *48*, 150–158. [CrossRef]
16. Cerecero, E.R. Etude des écoulements et des transferts thermiques lors de la fabrication d'un sorbet á l'échelle du pilote et du laboratoire. Ph.D. Thesis, L'institut National Agronomique Paris Grignon, Paris, France, 2003.
17. González-Ramírez, J.E.; Leducq, D.; Arellano, M.; Alvarez, G. Energy consumption optimization of a continuous ice cream process. *Energy Convers. Manag.* **2013**, *70*, 230–238. [CrossRef]
18. Kitanovski, A.; Vuarnoz, D.; Ata-Caesar, D.; Egolf, P.W.; Hansen, T.M.; Doetsch, C. The fluid dynamics of ice slurry. *Int. J. Refrig.* **2005**, *28*, 37–50. [CrossRef]
19. Benkhelifa, H.; Alvarez, G.; Flick, D. Development of a scraper-rheometer for food applications: Rheological calibration. *J. Food Eng.* **2008**, *85*, 426–434. [CrossRef]
20. Skelland, A.H.P.; Leung, L.S. Power consumption in scraped heat exchange. *Br. Chem. Eng.* **1962**, *7*, 264–267.
21. Trommelen, A.M.; Beek, W.J. The mechanism of power consumption in a Votator†-type scraped-surface heat exchanger. *Chem. Eng. Sci.* **1971**, *26*, 1977–1986. [CrossRef]
22. Trommelen, A.M.; Boerema, S. Power consumption in a scraped-surface heat exchager. *Trans. Instin. Chem. Eng.* **1966**, *44*, T329–T334.
23. Läuger, J. *Rheological Characterization of Ice Cream*; Anton Paar Germany GmbH: Ostfildern, Germany, 2003; Available online: https://www.anton-paar.com/mx-es/servicios-y-soporte/documentos/reometro-de-la-serie-mcr/ (accessed on 31 January 2020).
24. Hernández-Parra, O.D.; Plana-Fattori, A.; Alvarez, G.; Ndoye, F.T.; Benkhelifa, H.; Flick, D. Modeling flow and heat transfer in a scraped surface heat exchanger during the production of sorbet. *J. Food Eng.* **2018**, *221*, 54–69. [CrossRef]
25. Liu, S.; Hao, L.; Rao, Z.; Zhang, X. Experimental study on crystallization process and prediction for the latent heat of ice slurry generation based sodium chloride solution. *Appl. Energy* **2017**, *185*, 1948–1953. [CrossRef]
26. Qin, F.G.; Premathilaka, S.; Chen, X.D.; Free, K.W. The shaft torque change in a laboratory scraped surface heat exchanger used for making ice slurries. *Asia Pacific J. Chem. Eng.* **2007**, *2*, 618–630. [CrossRef]
27. Ali, S.; Baccar, M. Numerical study of hydrodynamic and thermal behaviors in a scraped surface heat exchanger with helical ribbons. *Appl. Therm. Eng.* **2017**, *111*, 1069–1082. [CrossRef]

28. Ndoye, F.T.; Hernandez-Parra, O.; Benkhelifa, H.; Alvarez, G.; Flick, D. Influence of operating conditions on residence time distributions in a scraped surface heat exchanger during aerated sorbet production. *J. Food Eng.* **2018**, *222*, 126–138. [CrossRef]

29. Guo, E.; Zeng, G.; Kazantsev, D.; Rockett, P.; Bent, J.; Kirkland, M.; Van Dalen, G.; Eastwood, D.S.; StJohn, D.; Lee, P.D. Synchrotron X-ray tomographic quantification of microstructural evolution in ice cream-a multi-phase soft solid. *RSC Adv.* **2017**, *7*, 15561–15573. [CrossRef]

30. Mo, J.; Groot, R.D.; McCartney, G.; Guo, E.; Bent, J.; van Dalen, G.; Schuetz, P.; Rockett, P.; Lee, P.D. Ice crystal coarsening in ice cream during cooling: A comparison of theory and experiment. *Crystals* **2019**, *9*, 321. [CrossRef]

31. Zhu, Z.; Zhou, Q.; Sun, D.W. Measuring and controlling ice crystallization in frozen foods: A review of recent developments. *Trends Food Sci. Technol.* **2019**, *90*, 13–25. [CrossRef]

32. Mo, J.; Guo, E.; McCartney, D.G.; Eastwood, D.S.; Bent, J.; Van Dalen, G.; Schuetz, P.; Rockett, P.; Lee, P.D. Time-resolved tomographic quantification of the microstructural evolution of ice cream. *Materials* **2018**, *11*, 2031. [CrossRef]

33. Dehkordi, K.S.; Fazilati, M.A.; Hajatzadeh, A. Surface Scraped Heat Exchanger for cooling Newtonian fluids and enhancing its heat transfer characteristics, a review and a numerical approach. *Appl. Therm. Eng.* **2015**, *87*, 56–65. [CrossRef]

34. Yataghene, M.; Pruvost, J.; Fayolle, F.; Legrand, J. CFD analysis of the flow pattern and local shear rate in a scraped surface heat exchanger. *Chem. Eng. Process. Process Intensif.* **2008**, *47*, 1550–1561. [CrossRef]

35. Steffe, J.F. *Rheological Methods in Foods Proccess Engineering*; Freeman Press: Eats Lansing, MI, USA, 1996.

36. Mabit, J.; Fayolle, F.; Legrand, J. Shear rates investigation in a scraped surface heat exchanger. *Chem. Eng. Sci.* **2003**, *58*, 4667–4679. [CrossRef]

37. Chen, J.; Ma, C.; Ji, X.; Lu, X.; Wang, C. Mechanism Study of Waste Heat Recovery from Slurry by Surface Scraped Heat Exchanger. *Energy Procedia* **2017**, *105*, 1109–1115. [CrossRef]

Article

Assessment of the Dynamics Flow Field of Port Plate Pair of an Axial Piston Pump

Lingxiao Quan [1,2]**, Haihai Gao** [1]**, Changhong Guo** [1,*] **and Shichao Che** [1]

[1] College of Mechanical Engineering, Yanshan University, Qinhuangdao 066004, China; lingxiao@ysu.edu.cn (L.Q.); ghh2427113346@163.com (H.G.); shichaoche@163.com (S.C.)

[2] Hebei Provincial Key Laboratory of Heavy Machinery Fluid Power Transmission and Control, Yanshan University, Qinhuangdao 066004, China

[*] Correspondence: guochanghong@ysu.edu.cn

Received: 19 November 2019; Accepted: 7 January 2020; Published: 8 January 2020

Abstract: This paper aims at studying the dynamic fluid evolution process of port plate pair of an axial piston pump. First of all, The Renormalization Group $k - \varepsilon$ model (RNG $k - \varepsilon$ model) is implemented to simulate the dynamic flow distribution and forecast the evolution of the internal vortex structure inside the valve plate chamber with different speeds of pistons and velocities of inlet fluid by using computational fluid dynamics software. Then, an equivalent amplification test model of a piston-valve plate is built up based on Reynolds similarity theory; the flow state of the piston-valve plate flow field is observed applied the particle image velocimetry (PIV) measuring technique. The resulting uniformity of numerical simulation and PIV measurement verifies that the RNG $k - \varepsilon$ model can achieve high-precision prediction for the vortex structure inside the valve plate chamber. Through analysis of velocity contours and streamlines of the flow field, it can be found that vortices with different scales, strengths and positions will occur during the process of fluid distribution, and the scale and strength of the vortex inside the valve plate chamber will be reduced with the increase of the piston's moving speed, so the energy loss is also reduced and the efficiency is improved.

Keywords: axial piston pump; RNG k-ε model; flow distribution characteristics; PIV measurements

---

1. Introduction

For high-end equipment hydraulic systems, the axial piston pump will turn towards the development of high-speed and high-pressure, which is in accordance with the development trend of a high power-to-weight ratio. However, as the speed and pressure increase, the fluid medium is in a state of high-speed rotation and three-dimensional, unsteady turbulent flows will appear inside the valve plate chamber, which will cause the unsteady flow structures and fluid-induced vibration to be more sophisticated in the energy transfer process of the axial piston pump. Moreover, the fluid vibration mechanism and vibration and noise reduction of pumps have attracted widespread study [1–6]. Therefore, it is crucial to accurately calculate and analyze the dynamic distribution process of fluid of the axial piston pump, which could provide a theoretical base for reducing the energy loss and improving energy-delivery efficiency of axial piston pump.

An enormous amount of researches have been carried out on the flow field of pumps either by the numerical simulation method or measuring techniques. For instance, through using the Zwart-Gerber-Belamri cavitation model to simulate the cavitating flow field in a centrifugal pump, the applicability of RNG $k - \varepsilon$ and LES turbulence model has been compared [7]. Xu et al. calculated the interaction in a pump based on three-dimensional Navier-Stockes equation and RNG $k - \varepsilon$ model, after creating an interface between the rotor and stator [8]. Zhang et al. examined the applicability of the standard $k - \varepsilon$ model, the renormalization group (RNG $k - \varepsilon$) model and the Realizable $k - \varepsilon$ model using numerical simulations of a centrifugal pump [9]. Wu et al. used four turbulent models to predict the performance of a single-channel pump at a large flow rate, the results showing that the RNG $k - \varepsilon$ model was the best one among four models: the standard $k - \varepsilon$, RNG $k - \varepsilon$ (renormalization group $k - \varepsilon$), standard $k - \omega$, and SST $k - \omega$ (shear stress transport $k - \omega$) [10]. Kumar S et al. studied the momentum exchange of fluid in the sliding block of a piston pump with a groove, and the vorticity in the groove was analyzed under different operating modes based on CFD simulation [11]. The flow pulsation characteristics of axial piston pump are studied when the cross angles are different to provides a theoretical basis for noise reduction of the pump by using CFD technology [12]. Guan C et al. studied the flow pulsation and instantaneous pressure characteristics of the piston chamber using computational fluid dynamics, which provided suggestions for the optimization design of the axial-piston pump in aviation [13]. The flow pulsation characteristics of an axial piston pump were researched by CFD technology, and the pump flow condition was tested [14].

Experimental measurement is a key means to study the inner flow law of turbo machinery. Particle Image Velocimetry (PIV) plays an important role in measurement, depending on its unparalleled advantage of non-contact measurement technology. For instance, Sinha et al. used PIV to research the transient flows and turbulence structures in a centrifugal pump [15]. Wuibaut studied the velocity fields of a radial flow pump through 2D PIV technology [16]. Wu et al. adopted PIV to measure the global flow structures in the model pump under different working conditions [17]. Keller et al. used PIV to measure the unsteady flow structures in a volute centrifugal pump with high flow rate [18]. Li et al. measured and analyzed the unsteady flow field in a mixed-flow pump on Particle Image Velocimetry (PIV) [19]. Zhou et al. validated different turbulence models used for numerical simulation of a centrifugal pump diffuser through PIV measuring technology [20]. The unsteady flow structure and its evolution in a low specific speed centrifugal pump were measured by employing PIV technology [21].

Although a large number of researchers have conducted a comprehensive study on the flow field of axial piston pump, there are still some deficiencies in the research about the assessment of the dynamics flow field of a port plate pair of an axial piston pump. In this respect, the innovation and contribution of this paper are as follows:

(1) After investigation and discussion, a high-precision turbulence model for assessment of the dynamics flow field of port plate pair in an axial piston pump is determined.

(2) Based on the Reynolds similarity criterion, the similarity calculation and experimental model design are carried out. An enlarged model of port plate pair of an axial piston pump was established, which is twice the size of the prototype model.

(3) Based on computational fluid dynamics (CFD) techniques and Particle Image Velocimetry (PIV), the dynamics flow field of the port plate pair is simulated and validated with a test when the speed of the piston and the flow of the pump inlet are different.

(4) Through the comparison of the results of CFD calculations and PIV measurements, the influence of piston speed and pump inlet flow on the dynamic characteristics of the flow field of the port plate pair is analyzed and the sources of errors are discussed.

2. Structure of Axial Piston Pump and Port Plate Pair

The basic structure of the axial piston pump is shown in Figure 1. Usually, the pump shaft is driven by an electric motor or engine. The cylinder block is connected and rotates with the pump shaft. Pistons rotate around the pump shaft and reciprocate along its axis direction owing to the tilt angle of the swash plate, which can allow periodic change of the volume of the piston chamber and realize oil suction and extrusion of the pump.

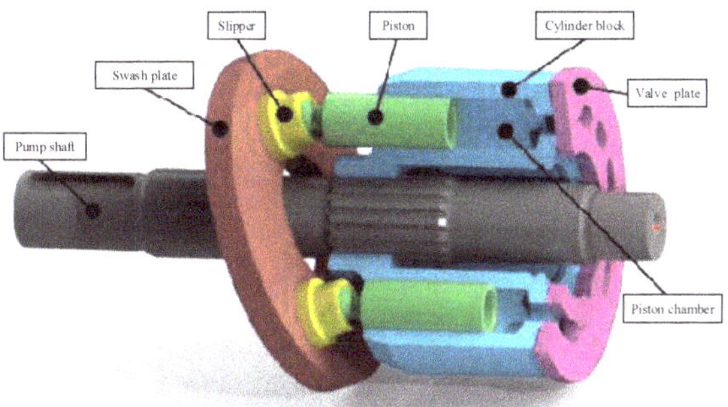

Figure 1. Structure of Swash-plate axial piston pump.

For the axial piston pump, the most complex region of the flow field is in the port plate pair consisting of the piston and valve plate. Figure 2 shows the port plate pair of the axial piston pump.

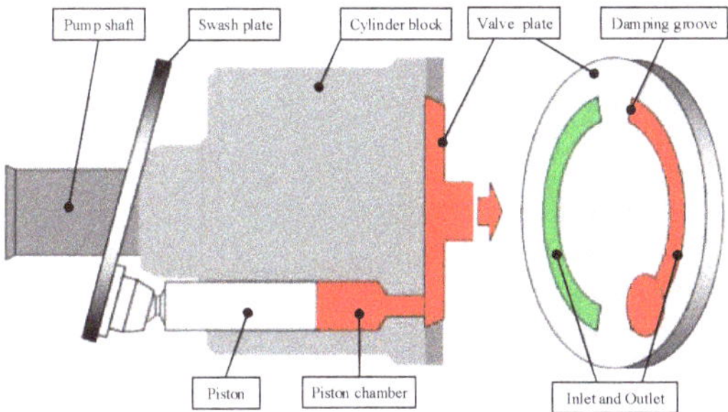

Figure 2. The port plate pair of axial piston pump.

The functions of the valve plate are to distribute and deliver inlet and outlet fluid, the fluid will go through a complex and variable flow channel in the distribution of the flow field, and the momentum and velocity of the fluid will undergo violent changes. This may give rise to the service life and energy delivery efficiency reduction of the axial piston pump as well as vibration, noise and other issues. Therefore, it is of great significant to study the dynamics flow field of the port plate pair of an axial piston pump distribution flow field of the pump accurately.

3. Turbulence Model

There are many types of flow channel structures in the pump; a universal turbulence model that is applicable for all flow problems has not yet been formulated, this is because the nature of turbulence is an irregular condition of flow, which has strong nonlinearity and large width features in the dimension of length, time, and velocity. An appropriate turbulence model is key to characterize the flow fields when employing computational fluid dynamics (CFD). The common used turbulence models are as follows: the direct numerical simulation (DNS), large eddy simulation (LES) and the Eddy Viscosity models [22]. DNS (direct numerical simulation): Navier-Stokes equations are numerically simulated at all lengths and all scales, so its computational requirements are far too high for any practical application. LES (large eddy simulation): The major vortices are analytically solved, while sub-grid eddies are described with sub-grid models. RANS (Reynolds averaged Navier-Stokes): The entire flow is averaged and the turbulence is modelled using various approaches, which can reduce the physical complexity of turbulent flow and increase the accuracy of the simulation [23]. An increase in orders of the turbulence model will result in a decrease in the number of assumptions according to the sequence of RSM (Reynolds Stress Model), LES (Large Eddy Simulation) and DNS (Direct Numerical Simulation).

The prediction accuracy increases greatly from $k-\varepsilon$ to DNS, while the demand on computational time and capability of computational facilities is higher [24]. The DNS and LES have high computation cost for most engineering applications, therefore, Reynolds-averaged Navier-Stokes (RANS) equations is an economic solution to turbulent flow [22].

The renormalization group (RNG) $k-\varepsilon$ turbulence model has its particular advantages in the simulation of turbulent fluid motion. The coefficients of the renormalization group (RNG) $k-\varepsilon$ turbulence model is derived from theory rather than the experimental fit method employed in the standard model and an additional strain term is introduced into the dissipation rate equation. The RNG $k-\varepsilon$ model makes a better prediction of the vortex structure and flow separation characteristics, because it could consider the strong anisotropy at regions where large shear is occurring. In addition, the swirling flow effect on the turbulence flow is also considered in the RNG $k-\varepsilon$ model, this provides further evidence that the RNG $k-\varepsilon$ model is fitter for the prediction of the large curvature and strain rate flow [25,26].

After a comprehensive consideration of the computing costs and the demand for analytical precision of the strong curvature flow field, the RNG $k-\varepsilon$ model is the most suitable for researching the distribution flow field.

The equations of the RNG $k-\varepsilon$ model are expressed as [26,27].

$$\rho\frac{\partial k}{\partial t} + \rho\frac{\partial}{\partial x_i}(ku_i) = \frac{\partial}{\partial x_j}\left[\left(\mu + \frac{\mu_t}{\sigma_k}\right)\frac{\partial k}{\partial x_j}\right] + G_k + G_b - \rho\varepsilon - Y_M \tag{1}$$

$$\rho\frac{\partial \varepsilon}{\partial t} + \rho\frac{\partial}{\partial x_i}(\varepsilon u_i) = \frac{\partial}{\partial x_j}\left[\left(\mu + \frac{\mu_t}{\sigma_\varepsilon}\right)\frac{\partial \varepsilon}{\partial x_j}\right] + C_{1\varepsilon}\frac{\varepsilon}{k}(G_k + C_{3\varepsilon}G_b) - C_{2\varepsilon}\rho\frac{\varepsilon^2}{k} - R \tag{2}$$

In Equations (1) and (2), the parameters are defined as follows: x_i is the i axis coordinate; x_j is the j axis coordinate; k is the turbulent kinetic energy; ε is the dissipation rate; ρ is the fluid density; u_i is the fluid velocity; u_t is the turbulent viscosity coefficient; μ is the dynamic viscosity of fluid; G_k is the turbulent kinetic energy generated by the average flow gradient; G_b is the turbulent kinetic energy caused by the buoyancy effect; Y_M is the influence of compressible turbulent fluctuation on the total dissipation rate; and R is the correction term. The coefficients of the RNG $k-\varepsilon$ model are described as: $C_{1\varepsilon} = 1.42$; $C_{2\varepsilon} = 1.68$; $C_{3\varepsilon} = 1.0$ and $C_\mu = 0.0845$. The turbulent Prandt number of k and ε can be expressed as: $\sigma_k = 1.0$ and $\sigma_\varepsilon = 1.0$.

4. Modeling and Numerical Simulation

4.1. Similarity Calculation of the Flow Field

Due to the smaller size and larger radius of curvature of the plunger cavity and the valve plate, there will be strong refraction and scattering phenomena in the pump. In order to facilitate flow field testing, the actual piston-valve plate model should be amplified equivalently because the fluid inside the pump is mainly affected by the fluid resistance and is mainly related to the viscous force of the fluid. Therefore, the equivalent amplification model of the piston-valve plate flow field was built according to the Reynolds similarity criterion. It is required that the Reynolds number of the experimental model is the same as that of the prototype model and the relationship between the parameters is shown as:

$$\frac{\rho_p l_p v_p}{\mu_p} = \frac{\rho_m l_m v_m}{\mu_m} = \mathrm{Re} \tag{3}$$

In Equation (3), the subscript p means the prototype model; the subscript m means the experimental model; ρ is the fluid density; l is the characteristic length; μ is the fluid dynamic viscosity; and v is the moving speed of the piston.

To facilitate the similarity calculation, the following definitions are made: k_ρ is the density ratio coefficient; k_v is the velocity ratio coefficient; k_l is the length ratio coefficient; and k_μ is the viscosity ratio coefficient.

$$k_\rho = \frac{\rho_p}{\rho_m} \tag{4}$$

$$k_v = \frac{v_p}{v_m} \tag{5}$$

$$k_l = \frac{l_p}{l_m} \tag{6}$$

$$k_\mu = \frac{\mu_p}{\mu_m} \tag{7}$$

Equations (4)–(7) are substituted into Equation (3) to get Equation (8) as follows:

$$\frac{k_\rho k_v k_l}{k_\mu} = 1 \tag{8}$$

According to the limitation of performance of the test system and the requirements of the experimental site, the piston-valve plate will be magnified to twice its actual size; this means the length ratio coefficient is 0.5.

The viscosity of the hydraulic oil in the actual pump is high and the tracer particles cannot be evenly distributed in the fluid. The density ratio coefficient and viscosity ratio coefficient can be obtained as $k_\rho = 0.87$ and $k_\mu = 45.91$, and $k_v = 105.54$, which can be calculated according to Equation (8). The density and viscosity parameters of the two fluid medium are shown in Table 1, when the temperature is 20 °C.

Table 1. The fluid parameters of the prototype and experimental model.

Parameters	l (mm)	ρ (kg/m^3)	μ (Pa·s)	v (mm/s)	Re
Prototype	16	870	0.046000	4241.15	1283.41
Experiment	32	1000	0.001002	40.18	1283.19
Proportion	0.5	0.87	45.91	105.55	1

4.2. The Piston-Valve Plate Model

It is important to note that the piston-valve plate model in the present study is designed to research the turbulence characteristics of the flow field in the flow distribution process of the axial piston pump. According to the motion law of the piston, the actual structure of the piston-valve plate and the scale ratio coefficient, the computational model is built as Figure 3a. It is composed of the piston chamber, valve plate and outlet; the area of the valve plate is set to be the testing area. As Figure 3b shows, four typical positions that the piston moves through the valve plate are selected for analysis.

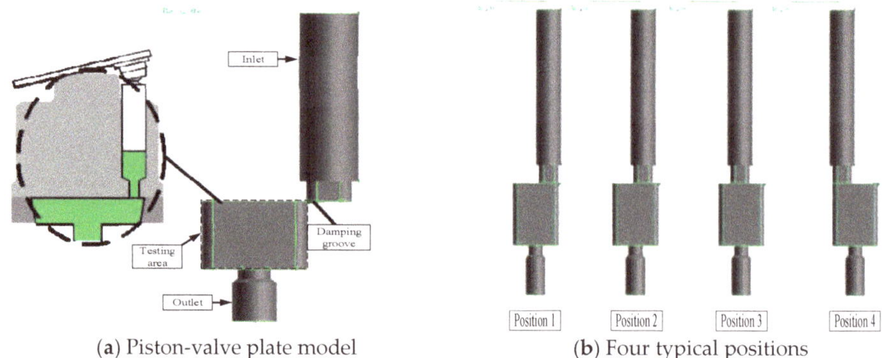

(**a**) Piston-valve plate model   (**b**) Four typical positions

Figure 3. Piston-valve plate model.

4.3. Computational Meshes and Computational Method

The quality of meshes is fundamental to the numerical simulation result. Figure 4 shows the piston-valve plate model and computational mesh model; the grid is generated by preprocessing software Gambit. The piston chamber and the outlet area are meshed with hexahedral grids; the structure of the valve plate area is meshed with tetrahedral grids. Due to the great pressure gradient and pressure changing rate during the pre-loading and pre-unloading, the mesh of the triangular damping groove is refined to improve the simulation accuracy. Considering the complexity of the flow field, the value between 10–500 can meet the requirements of calculation accuracy. The highest and lowest values of y+ approximately equal to 500 and 30 after a series of adjustments in the simulation. Grid density, as the basis of the evaluation, should be overall examined; three grid resolutions are tested, namely, coarse (1,762,423 cells), medium (2,945,787 cells) and fine (4,845,123 cells). Taking into account the efficiency and accuracy of simulation, medium mesh (2,945,787 cells) was finally selected for the following simulation analysis by FLUENT (version 16.0).

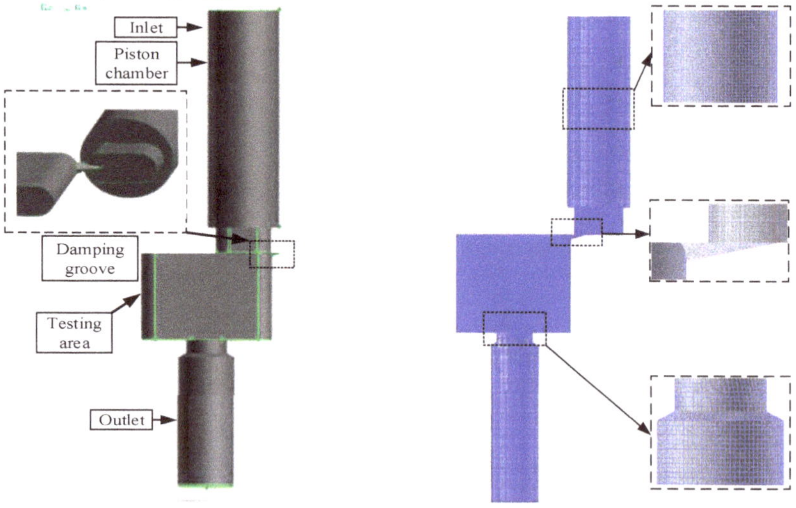

(**a**) Piston-valve plate model   (**b**) Computational domain and mesh

Figure 4. Piston-valve plate model and computational mesh model.

Due to the relative slip motion of the fluid between the piston chamber and valve plate, the transient interaction will occur between the two areas and so the sliding mesh technique is used to deal with the data exchange of the interface between the piston and valve plate.

Unsteady numerical investigation is carried out in the piston-valve plate of an axial piston pump by using the commercial CFD code ANSYS-FLUENT. A pressure-based solver is employed in the simulation, using SIMPLE method to couple pressure and velocity. FLUENT provides an optional discretization scheme for each governing equation. In order to achieve higher-order levels at cell faces, a second-order upwind scheme is adopted for the spatial discretization method of the computational domain. The steps in one numerical simulation cycle are set as 500, to better simulate a transient flow, it is essential to set the time step at least one order of magnitude less than the smallest time constant in the system. Observing the number of iterations required per time step to converge is a good way to judge its choice, the ideal number of iterations is 5–10 each time step. Accordingly, the time step is set to 0.0008 s.

4.4. Boundary Conditions

When analyzing the distribution flow field, the inlet boundary conditions and the outlet boundary conditions should be set. They are described as follows.

4.4.1. The Inlet Boundary Conditions

In order to study the impact of the piston moving speed and the inlet fluid velocity on the distribution flow field characteristics, a series of sets of piston moving speeds and pump inlet flows are set up. Table 2 shows the inlet boundary conditions.

Table 2. The inlet boundary conditions.

Sets 1	
Piston Moving Speed (mm/s)	**Pump Inlet Flow (mL/min)**
5.9	750
5.9	850
5.9	950
Sets 2	
Piston Moving Speed (mm/s)	**Pump Inlet Flow (mL/min)**
3.7	750
5.9	750
8.3	750
12.5	750

The inlet velocity is obtained by the pump inlet flow data divided by the cross sectional area of the piston chamber ($A = \pi d^2/4 = 803.84\,\text{mm}^2$, $d = 32$ mm, which is the diameter of the piston chamber). The application of MATLAB to fitting the inlet velocity curve for different pump inlet flows and piston moving speeds is shown in Figure 5.

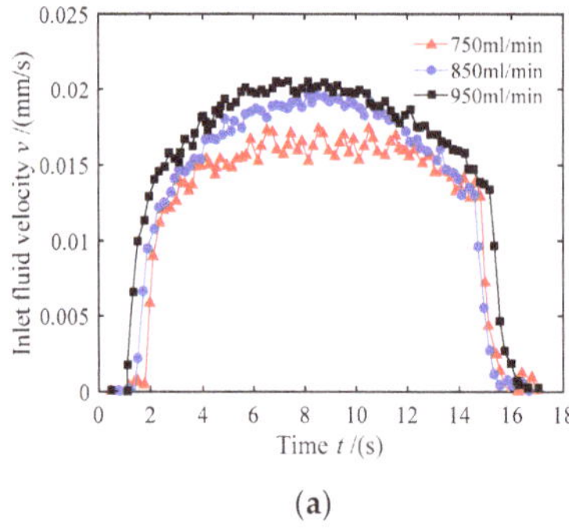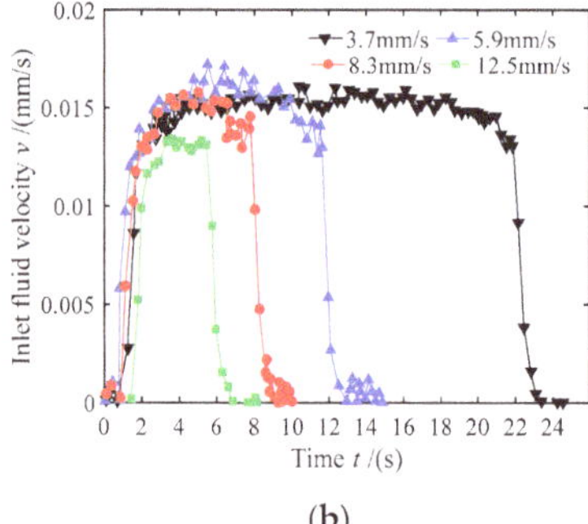

(a)   (b)

Figure 5. Inlet fluid velocity of piston chamber with different boundary conditions. (**a**) Inlet fluid velocity with different pump inlet flows; (**b**) Inlet fluid velocity with different piston moving speeds.

4.4.2. The Outlet Boundary Conditions

During the experiment, the outlet pipeline is directly connected to the water tank, so the outlet boundary condition is set as the constant pressure outlet with one standard atmosphere.

4.5. *Simulation of the Piston-Valve Plate Distribution Flow Field*

The simulation results are conducted with different pump inlet flows and different piston moving speeds.

4.5.1. Simulation Results of Dynamic Flow Field with Different Pump Inlet Flows

When the piston moving speed was 5.9 mm/s, the numerical simulation results of dynamic flow fields with different pump inlet flows are shown in Figures 6–10. Figure 6 presents the velocity contours and streamlines when the pump inlet flow was 750 mL/min.

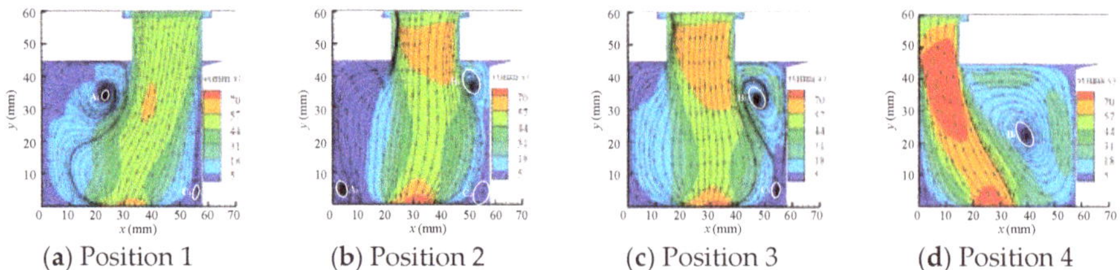

(**a**) Position 1   (**b**) Position 2   (**c**) Position 3   (**d**) Position 4

Figure 6. Velocity contours and streamlines with a pump inlet flow of 750 mL/min.

As can be observed in Figure 6a, as the piston initially moves along the negative direction of the *x*-axis, the throttle area gradually increases and the jet flow velocity also rises. The channel area will increase to closely match the cross-sectional area of the piston chamber and the jet flow velocity will be significantly greater than the piston moving speed. On both sides of the main flow, viscous fluid forms the boundary layer in the upper and right wall area of the valve plate chamber so the fluid will be separated and the clockwise vortex A and counter-clockwise vortex C will be generated.

From Figure 6a–d, it can be seen that the jet flow direction is gradually deflected with the movement of the piston chamber and, at last, the jet flow direction will be almost along the negative direction of the *y*-axis. Influenced by such a phenomenon, the vortex core of vortex A will gradually move down to the bottom of the valve plate and, during the process, the scale of vortex A will decrease gradually and dissipate eventually. At the same time, the cross sectional area of the triangular damping groove shrinks rapidly along the *x*-axis. Thus, the scale of vortex B increases continually, due to the throttling effect of the triangular damping groove. As the main flow moves along the negative direction of the *x*-axis, vortex C will move up continuously and the scale of it will gradually increase. At last, vortex C will merge with vortex B and a large vortex will be generated.

During the movement of the piston, the scale of vortex A is gradually decreasing and the scale of vortex B and vortex C are gradually increasing. Because vortex B is located near the triangular damping groove, it is easy to produce a larger fluid shear stress in this area, which will cause cavitation. Such a phenomenon will lead to damage and vibration of the valve plate.

Furthermore, due to the friction between the rotating fluid and the wall surface of the valve plate chamber and the viscous friction of the rotating fluid itself, the fluid kinetic energy is transformed into heat energy, which is the primary reason for the energy loss of the pump.

When the pump inlet flow was 850 mL/min and 950 mL/min, the velocity contours and streamlines of four typical positions are shown in Figures 7 and 8, respectively.

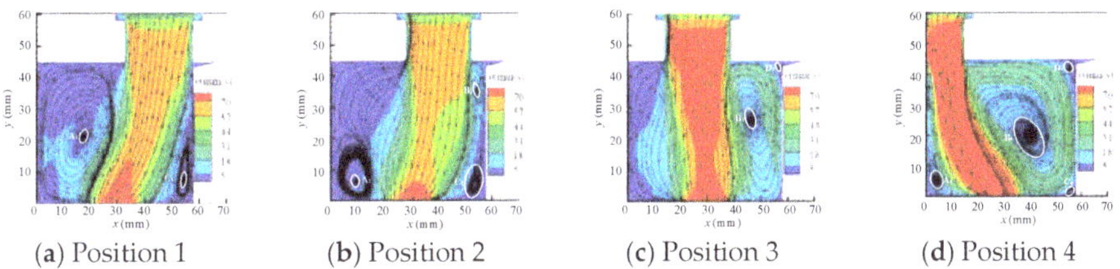

(**a**) Position 1   (**b**) Position 2   (**c**) Position 3   (**d**) Position 4

Figure 7. Velocity contours and streamlines with a pump inlet flow of 850 mL/min.

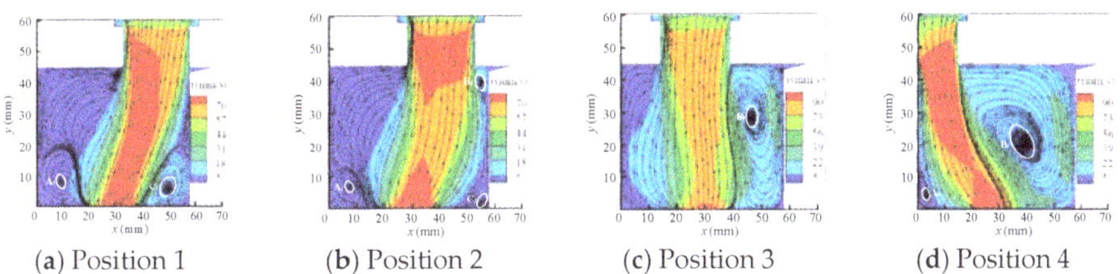

(**a**) Position 1   (**b**) Position 2   (**c**) Position 3   (**d**) Position 4

Figure 8. Velocity contours and streamlines with a pump inlet flow of 950 mL/min.

The conclusions drawn from Figures 7 and 8 are as follows:

1. The occurrence and development rule of the vortex system inside the valve plate do not change much with the increase of the pump inlet flow.
2. Comparison between Figures 6a, 7a and 8a shows that the core of vortex A moves down to the bottom of the valve plate chamber gradually as the pump inlet flow increases, meanwhile the scale of vortex A first becomes larger and then smaller. The scale of vortex C obviously increases with the increase of the pump inlet flow.
3. Comparison between Figures 6b, 7b and 8b shows that the core of vortex A moves down to the bottom of the valve plate chamber gradually with the increase of the pump inlet flow, while the scale of vortex A decreases considerably.
4. Comparison between Figures 6c, 7c and 8c shows that the core of vortex B moves down to a certain position and then no longer moves with the increase of the pump inlet flow, meanwhile the scale of vortex B increases to a certain extent and no longer changes.
5. Comparison between Figures 6d, 7d and 8d shows that the core of vortex B almost does not move with an increase in the pump inlet flow, but the scale of vortex B is slightly increased.

Generally speaking, the flow of the axial pump has little impact on the dynamic distribution when the rotating speed of the pump is constant.

4.5.2. Simulation Results of Dynamic Flow Field with Different Piston Moving Speeds

In order to research the influence of the piston moving speed on the flow field characteristics during the distribution process, four kinds of piston moving speeds were set up: 3.7 mm/s, 5.9 mm/s, 8.3 mm/s, and 12.5 mm/s.

The simulation results are shown in Figures 6 and 9, Figures 10 and 11.

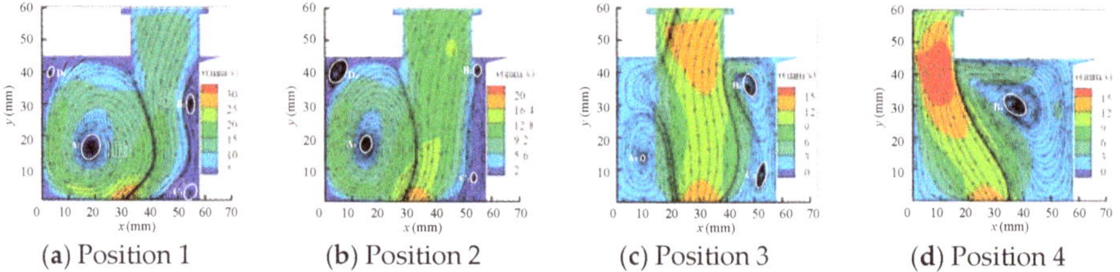

(**a**) Position 1   (**b**) Position 2   (**c**) Position 3   (**d**) Position 4

Figure 9. Velocity contours and streamlines with a piston moving speed of 3.7 mm/s.

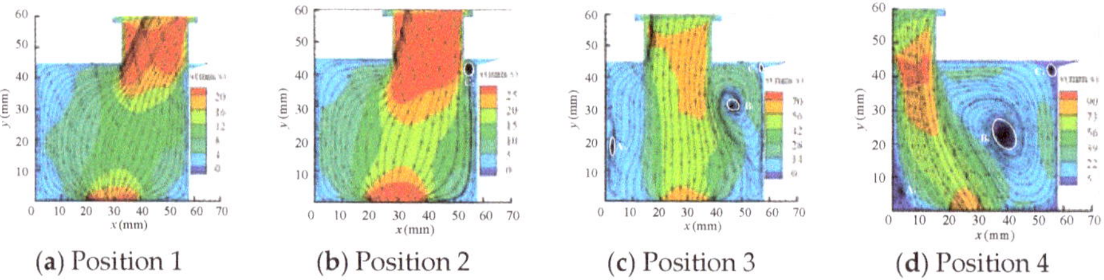

(**a**) Position 1   (**b**) Position 2   (**c**) Position 3   (**d**) Position 4

Figure 10. Velocity contours and streamlines with a piston moving speed of 8.3 mm/s.

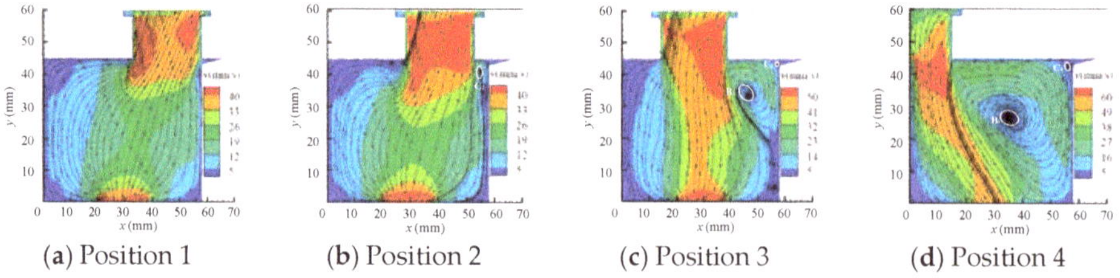

(**a**) Position 1   (**b**) Position 2   (**c**) Position 3   (**d**) Position 4

Figure 11. Velocity contours and streamlines with a piston moving speed of 12.5 mm/s.

The conclusions drawn from Figures 6 and 9, Figures 10 and 11 are as follows:

1. When the piston moves at a relatively low speed, such as 3.7 mm/s or 5.9 mm/s, the occurrence and development laws of the vortex system inside the valve plate are almost the same. When the piston moves at a speed of 3.7 mm/s, four vortexes come into being due to the fluid viscosity and the constraint of the valve plate wall surface, which are named as vortex A, vortex B, vortex C, and vortex D. It can be observed from picture b to picture d in Figures 6 and 9 that the core of vortex A moves toward the lower left as the piston chamber moves from the right to the left, meanwhile the scale of vortex A decreases until it disappears. At the same time, vortex B gradually moves down and combines with vortex C, which moves upward, until a large vortex is generated in the upper right of the valve plate chamber. However, there are also some differences between Figures 6 and 9. Vortex D will not appear in Figure 6 and vortex A disappears earlier (Position 3) when the piston chamber moving speed increases from 3.7 mm/s to 5.9 mm/s.

2. When the piston chamber moves at a relatively high speed, exceeding 8.3 mm/s, vortex D will not come into being and vortex C appears later (Position 2) than the low speed situations.

3. On the whole, the scale and strength of the vortex inside the valve plate chamber will reduce with an increase in piston moving speed, so the energy loss of the pump is also reduced and the efficiency is improved too. This means that improving the rotating speed of the pump does have the benefit of improving efficiency.

5. Experimental Analysis of the Dynamic Distribution Flow Field

5.1. Experiment Apparatus

The experimental test rig (Figure 12) includes three parts: the piston-valve plate slider system, the PIV testing system, and the data acquisition system. The piston chamber slider and the valve plate slider are manufactured out of acrylic materials with good light transmittance; the linear actuator is used to drive the piston chamber slider with different speeds.

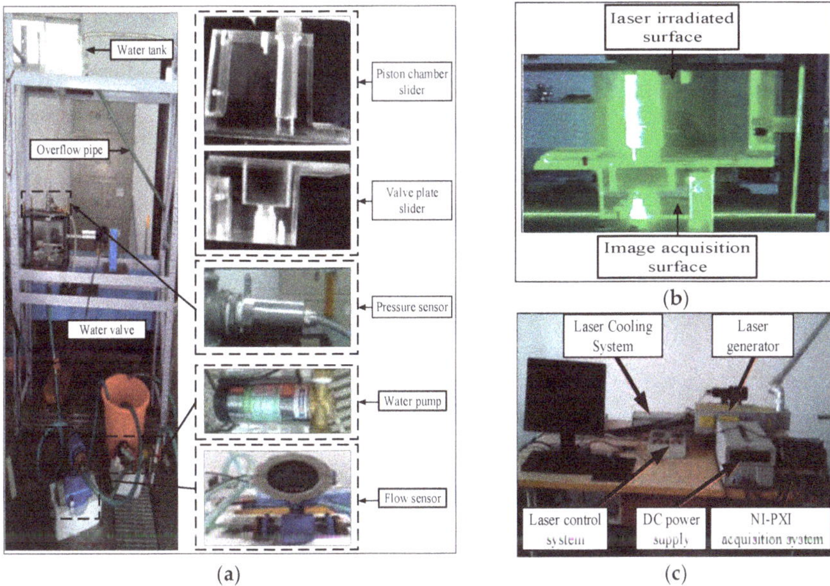

Figure 12. Experimental test rig. (**a**) Water supply system and piston-valve plate slider system; (**b**) PIV measuring areas; (**c**) PIV system and data acquisition system.

In order to capture the flow state of the piston-valve plate flow field, a 2D PIV system was applied, which is developed by Dantec Inc. (Bristol, UK). The PIV optical system (Figure 12c) can provide a sheet laser with a high brightness. The laser irradiated surface and the image acquisition surface are shown in Figure 12b.

During the experiment, the water pump provides water doped with tracer particles to the water tank. The water tank has an overflow opening to ensure that the piston chamber has a stable inlet pressure. The measurement plane of the flow is located at the pump outlet in the PIV test, by adjusting the opening of the water valve in the outlet pipeline; the maximum flow at the pump inlet is 750 mL/min, 850 mL/min, and 950 mL/min respectively, and the flow at the pump inlet is measured by the flow sensor.

The digital images were taken on a Digital CCD Camera with a resolution of 1600 Pixels × 1200 Pixels and 60 mm optical lens (Nikon Nikkor r 60/2.8). The NI-PXI data acquisition equipment is applied to collect the real-time data. Then, the NI-PXI data acquisition equipment is applied to collect the real-time data; the Dynamic Studio software stores the particle image data in the computer, analyzes and displays the velocity vector field in real time.

5.2. Experiments on Dynamic Flow Field with Different Pump Inlet Flows

During the experiment, the piston moving speed was set to be 5.9 mm/s and the pump inlet flows were set at 750 mL/min, 850 mL/min and 950 mL/min. The test results of the piston-valve plate dynamic distribution flow field are obtained during the process of the piston chamber slider moving from the left to the right of the valve plate slider. The results are shown as Figures 13–15.

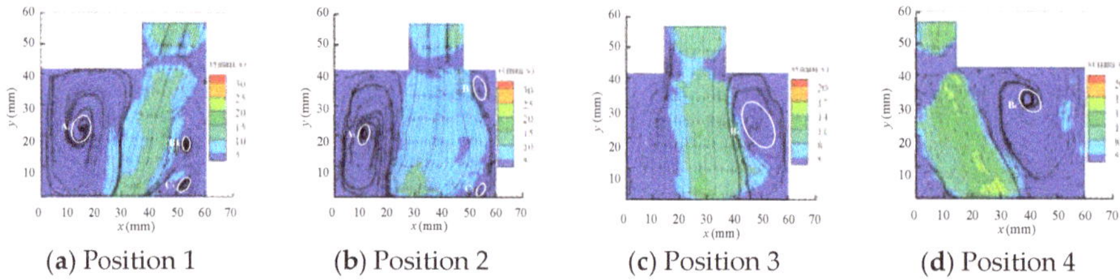

(**a**) Position 1   (**b**) Position 2   (**c**) Position 3   (**d**) Position 4

Figure 13. Velocity contours and streamlines with a pump inlet flow of 750 mL/min.

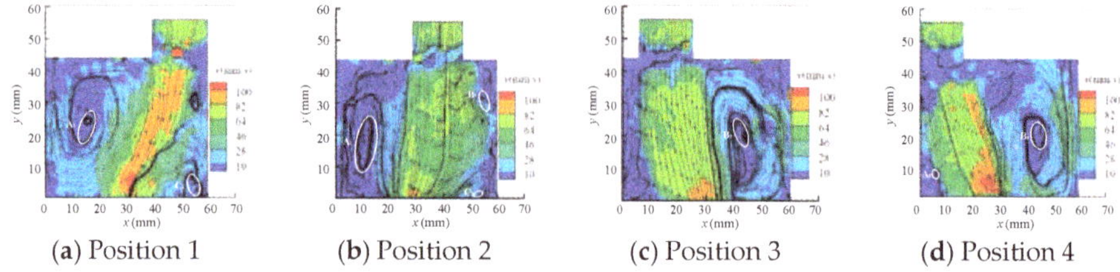

(**a**) Position 1   (**b**) Position 2   (**c**) Position 3   (**d**) Position 4

Figure 14. Velocity contours and streamlines with a pump inlet flow of 850 mL/min.

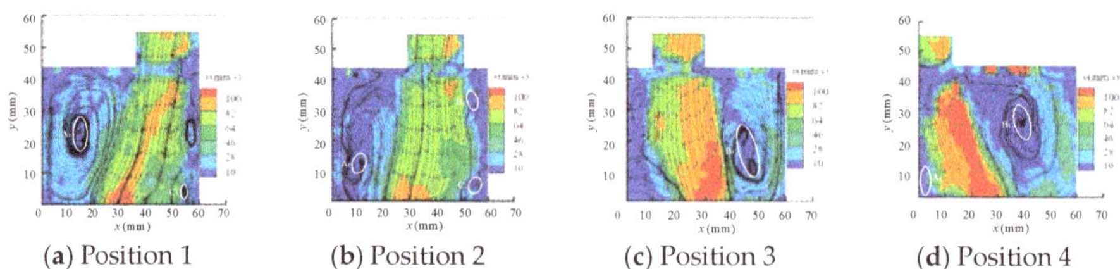

(**a**) Position 1   (**b**) Position 2   (**c**) Position 3   (**d**) Position 4

Figure 15. Velocity contours and streamlines with a pump inlet flow of 950 mL/min.

From the experimental results, it can be observed that there are three vortices generated inside the valve plate chamber, named vortex A, vortex B and vortex C, corresponding to the vortices from the numerical results in Figures 6–8. Furthermore, the comparison between the experimental results and the numerical results shows good consistency on the whole, that means the simulation results are reliable. However, there are some errors, such as vortex D in the simulation results had not been observed in the experiment. The reasons for it could be concluded into several items.

(1)   Manufacturing error: There is a guide groove structure between the piston chamber slider and the valve plate slider, its function is to ensure smoothness and stability of the relative sliding movement between the two sliders as shown in Figure 16. However, the material is thicker than in other regions and the laser cannot penetrate the whole structure, resulting in the PIV image details being blurred.

(2) Tracing particles: The concentration and distribution of the tracing particles have a slight effect on the accuracy of the experimental results.

(3) Bubbles: High-speed bubbles in the fluid mixing with tracer particles will cause chaos in the velocity contours and streamlines.

(4) Other reasons: Accuracy of test instruments, the refraction of lase, control error of piston speed, etc., will all lead to random errors in the experiment.

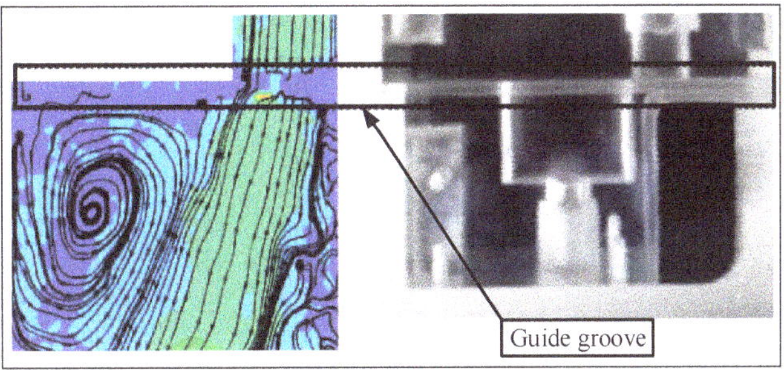

Figure 16. Guide groove.

5.3. Experiments on Dynamic Flow Field with Different Piston Moving Speeds

In the experiments, the piston moving speed was set to 3.7 mm/s, 5.9 mm/s, 8.3 mm/s, and 12.5 mm/s and the pump inlet flow was set to be 750 mL/min. The test results of the piston-valve plate dynamic distribution flow field were obtained during the process of the piston chamber slider moving from the left to the right of the valve plate slider. The results are shown in Figures 17–19.

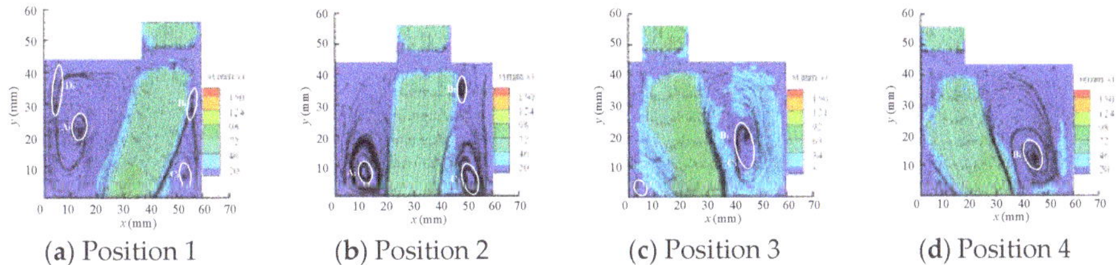

(**a**) Position 1 (**b**) Position 2 (**c**) Position 3 (**d**) Position 4

Figure 17. Velocity contours and streamlines with a piston moving speed of 3.7 mm/s.

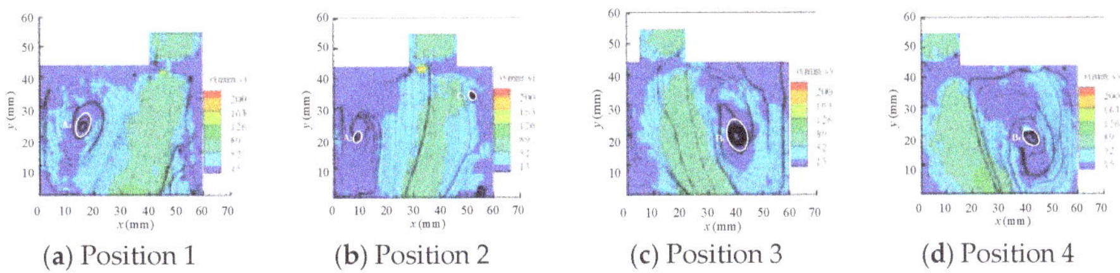

(**a**) Position 1 (**b**) Position 2 (**c**) Position 3 (**d**) Position 4

Figure 18. Velocity contours and streamlines with a piston moving speed of 8.3 mm/s.

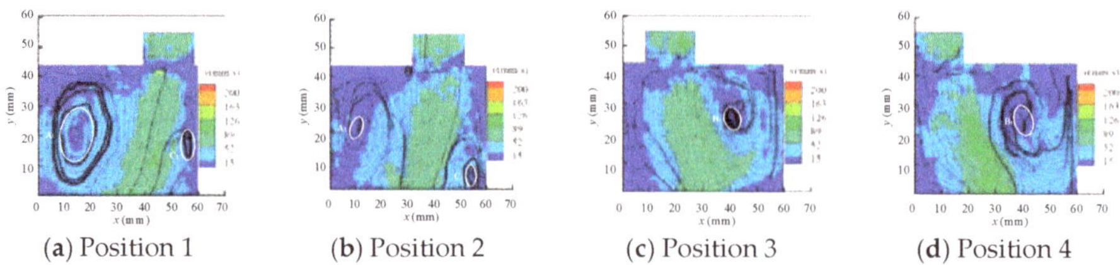

(**a**) Position 1   (**b**) Position 2   (**c**) Position 3   (**d**) Position 4

Figure 19. Velocity contours and streamlines with a piston moving speed of 12.5 mm/s.

As can be observed from the comparison between Figures 9–11 and Figures 17–19, the experimental results are basically consistent with the simulation results.

6. Conclusions

Based on the RNG $k - \varepsilon$ model and Reynolds similarity criterion, the dynamics flow field of the port plate pair of an axial piston pump is studied by both CFD calculations and PIV measuring technique. Simulations and measures are carried out at different speeds of the piston and flows of pump inlet; emphasis is placed on the dynamic fluid evolution process of the region of the piston valve plate; contour plots of fluid velocities and streamlines were presented and analyzed. The final numerical and experimental results allow to present the following general conclusions:

(1) The consistency of the numerical and experimental results shows that the RNG $k - \varepsilon$ model could realize an accurate analysis of flow field structures of a port plate pair in an axial piston pump.

(2) The results show that significant vortices with different scales, strengths and positions will occur in the process of distribution, which is closely associated with the speed of the piston and the flow rate of the pump inlet. The origin of this phenomenon is the throttling effect of the triangular damping groove and viscous effects of fluids.

(3) As a general observation, when the flow rate varies but the moving speed of the piston is constant, the occurrence and development law of vortex structures does not vary widely. However, large-sized vortices will appear near the triangular damping groove, which will lead to a large fluid shear stress, thus cavitation appears easily in the area, resulting in damage and vibration to the valve plate.

(4) When the piston moves at relatively low speeds, it is essentially the same law of the occurrence and development of vortex structures. However, when the piston moves at relatively high speeds, the scale of certain vortices reducted and some even disappeared, besides, some vortices appear later than that of low speeds. This indicates that the scale and strength of the vortices inside the valve plate chamber will reduce with an increase in piston moving speed. That is, enhancing the rotor speed does contribute to reduce the energy loss and improve the efficiency of pumps.

This paper supplies a theoretical basis for a better understanding of the dynamics flow field of a port plate pair of an axial piston pump. There are still shortcomings in the research. The actual flow field structure and piston motion of the axial piston pump are very complicated. The simulation and experimental results obtained by CFD techniques and the existing PIV test conditions are not completely consistent with the actual flow field distribution. In the future, the model's sphere of application will be extended by taking the time-varying parameters and the non-linear parameters into consideration; better and deeper research of the full three-dimensional flow field within the axial piston pump will be carried out.

Author Contributions: Conceptualization, L.Q., H.G. and C.G.; methodology, L.Q.; software, H.G.; validation, L.Q., H.G. and S.C.; formal analysis, H.G.; investigation, H.G.; resources, L.Q.; data curation, H.G.; writing—original draft preparation, H.G.; writing—review and editing, H.G.; visualization, H.G.; supervision, L.Q.; project administration, C.G.; funding acquisition, L.Q. All authors have read and agreed to the published version of the manuscript.

Funding: This research was funded by the National Key Research and Development Program of China, grant number (2014CB046400) and the National Natural Science Foundation of China, grant number (51775477 and 51505410).

Acknowledgments: The authors gratefully acknowledge the support of the above fundings and the authors also thank China Scholarship Council for supporting a two-years research stay of the first author and the corresponding author at the RWTH Aachen University and Washington State University.

Conflicts of Interest: The authors declare no conflict of interest.

References

1. Wieczorek, U.; Ivantysynova, M. Computer Aided Optimization of Bearing and Sealing Gaps in Hydrostatic Machines—The Simulation Tool Caspar. *Int. J. Fluid Power* **2002**, *3*, 7–20. [CrossRef]
2. Deng, B.; Liu, X.H.; Wang, J.N.; Jian, K. Dynamic Simulation of the Flow Fluctuation in Water Hydraulic Axial Piston Pump. *Chin. Hydraul. Pneum.* **2004**, *336*, 1094–1096.
3. Yu, Z.; Sanbao, H.; Yunqing, Z.; Liping, C. Optimization and Analysis of Centrifugal Pump considering Fluid-Structure Interaction. *Sci. World J.* **2014**, *2014*, 1–9.
4. Casartelli, E.; Mangani, L.; Ryan, O.; Schmid, A. Application of transient CFD-procedures for S-shape computation in pump-turbines with and without FSI. *IOP Conf. Ser. Earth Environ. Sci.* **2016**, *49*. [CrossRef]
5. Van Rijswick, R.; Talmon, A.; van Rhee, C. Fluid structure interaction (FSI) in piston diaphragm pumps. *Can. J. Chem. Eng.* **2016**, *94*, 1116–1126. [CrossRef]
6. Ye, S.-G.; Zhang, J.-H.; Xu, B. Noise Reduction of an Axial Piston Pump by Valve Plate Optimization. *Chin. J. Mech. Eng.* **2018**, *31*, 85–100. [CrossRef]
7. He, M.; Guo, Q.; Zhou, L.J.; Wang, Z.W.; Wang, X. Application of two turbulence models for computation of cavitating flows in a centrifugal pump. *IOP Conf. Ser. Mater. Sci. Eng.* **2013**, *52*. [CrossRef]
8. Xu, Z.H.; Wu, Y.L.; Chen, N.X.; Liu, Y.; Wu, Y.Z. Simulation of turbulent flow in pump based on sliding mesh and RNG k-ε model. *J. Eng. Thermophys.* **2005**, *26*, 66–68.
9. Zhang, S. Applicability of k-ε Eddy Viscosity Turbulence Models on Numerical Simulation of Centrifugal Pump. *J. Mech. Eng.* **2009**, *45*, 238. [CrossRef]
10. Wu, X.; Feng, J.; Liu, H.; Ding, J.; Chen, H. Performance prediction of single-channel centrifugal pump with steady and unsteady calculation and working condition adaptability for turbulence model. *Nongye Gongcheng Xuebao/Trans. Chin. Soc. Agric. Eng.* **2017**, *33*, 85–91.
11. Kumar, S.; Bergada, J.M.; Watton, J. Axial piston pump grooved slipper analysis by CFD simulation of three-dimensional NVS equation in cylindrical coordinates. *Comput. Fluids* **2009**, *38*, 648–663. [CrossRef]
12. Wei, X.Y.; Wang, H.Y. The Influence of Cross Angle on the Flow Ripple of Axial Piston Pumps by CFD Simulation. *Appl. Mech. Mater.* **2012**, *220–223*, 1675–1678. [CrossRef]
13. Guan, C.; Jiao, Z.; Shouzhan, H.E. Theoretical study of flow ripple for an aviation axial-piston pump with damping holes in valve plate. *Chin. J. Aeronaut.* **2014**, *27*, 169–181. [CrossRef]
14. Zhang, B.; Ma, J.; Hong, H.; Yang, H.; Fang, Y. Analysis of the flow dynamics characteristics of an axial piston pump based on the computational fluid dynamics method. *Eng. Appl. Comput. Fluid Mech.* **2017**, *11*, 86–95. [CrossRef]
15. Sinha, M. Quantitative Visualization of the Flow in a Centrifugal Pump with Diffuser Vanes, Part I: On Flow Structures and Turbulence. *J. Fluids Eng.* **2000**, *122*, 97–107. [CrossRef]
16. Wuibaut, G. PIV Measurements in the Impeller and the Vaneless Diffuser of a Radial Flow Pump in Design and off-Design Operating Conditions. *J. Fluids Eng.* **2002**, *124*, 791–797. [CrossRef]
17. Wu, Y.L.; Liu, S.H.; Yuan, H.J.; Shao, J. PIV measurement on internal instantaneous flows of a centrifugal pump. *Sci. China Technol. Sci.* **2011**, *54*, 270–276. [CrossRef]
18. Keller, J.; Blanco, E.; Barrio, R.; Parrondo, J. PIV measurements of the unsteady flow structures in a volute centrifugal pump at a high flow rate. *Exp. Fluids* **2014**, *55*, 1820. [CrossRef]
19. Li, W.; Zhou, L.; Shi, W.-d.; Ji, L.; Yang, Y.; Zhao, X. PIV experiment of the unsteady flow field in mixed-flow pump under part loading condition. *Exp. Therm. Fluid Sci.* **2017**, *83*, 191–199. [CrossRef]
20. Zhou, L.; Bai, L.; Li, W.; Shi, W.; Wang, C.; de Souza Neto, E. PIV validation of different turbulence models used for numerical simulation of a centrifugal pump diffuser. *Eng. Comput.* **2018**, *35*, 2–17. [CrossRef]

21. Zhang, N.; Gao, B.; Li, Z.; Ni, D.; Jiang, Q. Unsteady flow structure and its evolution in a low specific speed centrifugal pump measured by PIV. *Exp. Therm. Fluid Sci.* **2018**, *97*, 133–144. [CrossRef]

22. Wu, B. CFD investigation of turbulence models for mechanical agitation of non-Newtonian fluids in anaerobic digesters. *Water Res.* **2011**, *45*, 2082–2094. [CrossRef] [PubMed]

23. Lorenzon, A.; Antonello, M.; Berto, F. Critical review of turbulence models for CFD for fatigue analysis in large steel structures. *Fatigue Fract. Eng. Mater. Struct.* **2018**, *41*, 762–775. [CrossRef]

24. Khan, Z.; Bhusare, V.H.; Joshi, J.B. Comparison of turbulence models for bubble column reactors. *Chem. Eng. Sci.* **2017**, *164*, 34–52. [CrossRef]

25. Cheng, Q.; Wu, W.; Li, H.; Zhang, G.; Li, B. CFD study of the influence of laying hen geometry, distribution and weight on airflow resistance. *Comput. Electron. Agric.* **2018**, *144*, 181–189. [CrossRef]

26. Yang, Y.; Li, J.; Wang, S.; Wen, C. Gas-liquid two-phase flow behavior in terrain-inclined pipelines for gathering transport system of wet natural gas. *Int. J. Press. Vessel. Pip.* **2018**, *162*, 52–58. [CrossRef]

27. Yakhot, V.; Orszag, S.A. Renormalization group analysis of turbulence. I. Basic theory. *J. Sci. Comput.* **1986**, *1*, 3–51. [CrossRef]

Article

Phase Change and Heat Transfer Characteristics of a Corrugated Plate Heat Exchanger

Huashan Su, Chaoqun Hu *, **Zhenjun Gao ***, **Tao Hu, Gang Wang and Wan Yu**

College of Mechanical & Power Engineering, China Three Gorges University, Yichang 443002, China; suhuashan@ctgu.edu.cn (H.S.); hutao042__@163.com (T.H.); gwang2019@126.com (G.W.); yuwan@ctgu.edu.cn (W.Y.)

* Correspondence: outstandinghu@126.com (C.H.); sdrzjx2005@163.com (Z.G.)

Received: 19 November 2019; Accepted: 9 December 2019; Published: 24 December 2019

Abstract: In order to reveal the evolution law of heat transfer during phase change in a corrugated plate flow passage of a plate heat exchanger, a two-dimensional two-channel model was established to simulate the process of heat transfer during phase change in an unsteady flow passage. The results show that when the time was <3/5T, the average Nusselt number and average heat flux of the heat exchange wall surface decreased with time, the average temperature of the cold fluid outlet increased, the average temperature of the hot fluid outlet decreased, and the volume fraction of the gas phase was no higher than 0.2. When the time was >3/5T, the average Nusselt number and the average heat flux of the heat exchange wall, as well as the outlet average temperature of the cold and hot fluid, reached stability, while the volume fraction of the gas phase increased rapidly. During the whole heat transfer process, the change in Nusselt number and heat flux along the heat transfer wall surface was basically the same, and its value fluctuated along the wall surface, displaying extrema at the exit, entrance, and corrugated corner. The temperature of the heat exchange wall fluctuated and increased along the Y-axis, and began stabilizing after a time >3/5T. As time went on, the temperature gradient of the hot and cold fluid outlet and the temperature difference between the two sides of the heat exchange wall decreased, whereas the relative parameters of the heat flow inlet section of the corrugated passage reached stability before those of the cold flow inlet section. The whole simulation process fully reflects the heat transfer mechanism during phase change in a corrugated plate flow passage of a plate heat exchanger.

Keywords: plate heat exchanger; numerical simulation; phase change; multiphase flow; heat transfer

---

1. Introduction

The plate heat exchanger (PHE) was first introduced in the dairy industry at the end of the 19th century [1]. Later, the plate heat exchanger was improved in terms of plate design and seal, so that it could be successfully applied in many industries for heating, refrigeration, and air conditioning and food processing, as well as in the chemical and marine industry, and for energy generation systems [2,3]. The plate heat exchanger structure is highly compact, allowing turbulence to form easily, and the heat transfer requires a larger surface area compared to other types of heat exchanger. It can be used in high-temperature and high-pressure environments, such as for evaporator heating, condenser refrigeration, and waste heat recovery [4–14].

Evaporative heat transfer in a PHE is caused by nucleate boiling and forced convection boiling, and the heat transfer coefficient is related to both. Due to the limited data on boiling heat transfer in a PHE, it is not clear which boiling mechanism is dominant. Grabenstein [15] presented a summary of single-phase flows in a PHE, concluding the correlation of two-phase flows based on a large number of experimental data, including R22 and ammonia as the working fluids. Manglik et al. [16] provided a material library for the design of single-phase and two-phase PHEs. Khan et al. [17] established a set of correlation equations for a two-phase evaporator with ammonia as the working medium, and they verified the influence of the chevron angle on the thermal and hydraulic performance of the evaporator. Li et al. [18] simulated the heat transfer during phase change and flow of a herringbone corrugated PHE, and they compared the results with single-phase flow under the same corrugated parameters. They found that the heat transfer coefficient of phase change flow was increased by 20–100%, and a greater inclination angle of the corrugated plate led to an enhanced heat transfer effect. Huang et al. [19] used R134a and R507a as working media and conducted experiments in a PHE with three different geometric structures, showing that the heat flux had a greater influence on the heat transfer coefficient and a smaller influence on the pressure drop. Parameters such as mass flow rate, vapor dryness, and ripple angle have little effect on the heat transfer but great influence on the pressure drop.

Similar to evaporation, condensation in a plate heat exchanger is a function of mass, mass flow rate, heat flux, fluid properties, plate surface geometry, and local flow state. Longo [20–24] established a numerical model based on 338 experimental data, and he obtained two kinds of correlation relationships between the equivalent Reynolds number below 1600 and the equivalent Reynolds number above 1600 for forced convection condensation. The model was compared with 516 experimental data, and the absolute average percentage deviation was 16%. Han et al. [25] conducted experiments on R410A and R22, and the angles of the corrugated herringbone plate were 45°, 35°, and 20°. Unlike the previous correlation, they included the influence of plate geometry, and the conclusion showed that the heat transfer coefficient and pressure drop increased with the increase in mass flow and steam mass, while they decreased with the decrease in saturation temperature and chevron angle. Mancin et al. [26,27] studied the partial condensation of R410A and R407C in two PHE geometric structures with different aspect ratios and channel numbers. The experimental results showed that the heat transfer coefficient increased with the increase in steam mass, while it decreased with the increase in saturation temperature difference.

The phase change mechanism of a plate heat exchanger is relatively complex, and many problems in the process of flow heat transfer still need to be further explored [28]. At present, studies are mainly carried out using experimental methods, while numerical simulations mostly involve single-phase flow heat transfer. Further exploration is needed to study the heat transfer characteristics of a PHE when gas–liquid phase change occurs. In this paper, the process of heat transfer during phase change in a herringbone corrugated plate heat transfer channel was numerically simulated, and the User Defined Feature (UDF) program was imported into the Volume of Fluid (VOF) model to calculate the mass transfer and energy transfer between the gas and liquid. The change law of fluid temperature, phase volume distribution, Nusselt number, heat flux, and heat transfer wall parameters were studied to reveal the mechanism of heat transfer during phase change in the PHE corrugated passage, which can provide some reference for the design and optimization of a plate heat exchanger.

2. Calculation Model

The section introduces the establishment of the physical model of corrugated channel of herringbone plate heat exchanger, the division of grid, the selection of mathematical model and boundary conditions, and the verification of grid independence and model verification. It is prepared for the discussion of numerical simulation.

2.1. Physical Model and Mesh Generation

The flow in a PHE channel is extremely complex. In order to reflect the internal heat transfer during phase change state, the three dimensional flow problem is reduced into a two dimensional flow problem. The number of corrugated plates in the PHE channel is large, and the three-section corrugated unit double-channel structure is adopted to reflect the flow heat transfer in the whole heat exchanger channel. The first unit represents the inlet part, the second unit represents heat transfer part of the intermediate flow, and the third unit represents the outlet part. The main geometric parameters of the plate [29] include the spacing of the ripples λ, the height of the ripples H and the Angle of the ripples β, which are reflected in the two-dimensional structure as the spacing of the ripples λ and the height of the ripples H. ICEM software is used to establish a two-dimensional geometric model and grid division, the selected geometric size is $\lambda = 8$ mm, $H = 4$ mm. The upper flow channel is a cold fluid, the lower flow channel is the hot fluid, and the middle of the upper and lower flow channel is the heat exchanger plate. The extension of the entrance and exit is the transition section. The model is shown in Figure 1.

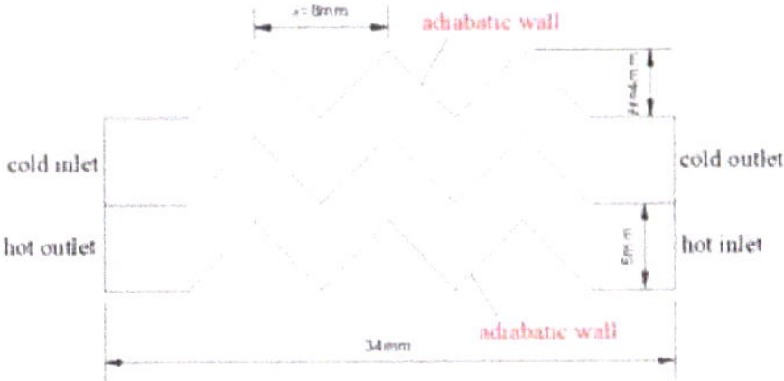

Figure 1. Physical model.

We use a structured grid to divide the model to get high quality grid. Due to the large temperature and velocity gradient at the near wall surface, the grid is refined at the near wall surface [30]. The grid division is shown in Figure 2. In the pre-processing software ICEM, the parameter used to evaluate the grid quality is "determinant $2 \times 2 \times 2$", and the display quality is 1, so the grid quality is the best. In order to ensure the accuracy of numerical simulation, the grid independence is verified. Figure 3 shows the change curve of $\overline{Nu}$ with the increase of mesh number. When the number of grids is 93,412, the value of $\overline{Nu}$ basically remains unchanged, indicating that the number of grids meets the requirements of calculation accuracy. Therefore, 93,412 is selected as the number of grids used for calculation.

Figure 2. Meshing generation.

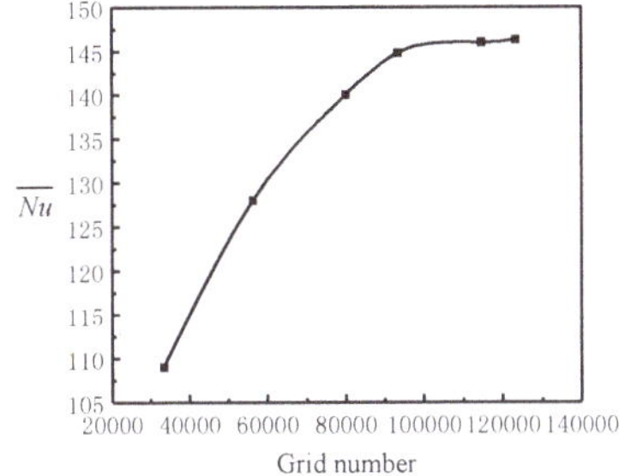

Figure 3. Grid independence verification.

2.2. Mathematical Model

In the calculation of multiphase flow, the heat and mass transfer model proposed by Schepper [31] is adopted to realize the gas-liquid phase change process between fluids. The mathematical model of the heat transfer during phase change process in plate heat exchanger was established.

Hypothesis: *(1) fluid is incompressible; (2) ignoring gravity and surface tension; (3) the viscous dissipative heat effect when fluid flow is ignored. The mathematical model equation is as follows:*

(1) Governing equation of heat and mass transfer

Mass transfer equations:
Evaporation:

$$S_M = \beta \alpha_l \rho_l \frac{(T_l - T_{sat})}{T_{sat}} \tag{1}$$

Condensation:

$$S_M = \beta \alpha_v \rho_v \frac{(T_{sat} - T_v)}{T_{sat}} \tag{2}$$

Energy transfer equations:
Evaporation:

$$S_E = \dot{m}_{lv} \Delta H = \beta \alpha_l \rho_l \frac{(T_l - T_{sat})}{T_{sat}} \Delta H \tag{3}$$

Condensation:

$$S_E = \dot{m}_{vl} \Delta H = \beta \alpha_v \rho_v \frac{(T_{sat} - T_v)}{T_{sat}} \Delta H \tag{4}$$

where S_M—the mass source term of evaporation and condensation, S_E—energy source term, α_v—vapor phase volume fraction, ρ_v—vapor phase density, α_l—liquid phase volume fraction, ρ_l—liquid phase density, ΔH—enthalpy, u—vapor phase velocity, T_l—liquid phase temperature, T_{sat}—phase transition temperature, T_v—vapor phase temperature, β—relaxation factor.

(2) Three conservation equations

Mass conservation equation:

$$\frac{\partial}{\partial t}(\alpha_v \rho_v) + \nabla \cdot (\alpha_v \rho_v u) = S_M \tag{5}$$

Momentum conservation equation:

$$\frac{\partial \rho u}{\partial t} + \nabla \cdot \left(\rho u u^T\right) = -\nabla \cdot p + \nabla \cdot \left(\mu\left(\nabla u + \nabla u^T\right)\right) + F \tag{6}$$

where ρ—density, μ—dynamic viscosity; ρ and μ depend on the volume fraction of all phases.
Energy conservation equation

$$\frac{\partial \rho E}{\partial t} + \nabla \cdot (u(\rho E + p)) = -\nabla \cdot p + \nabla \cdot \left(k_{eff} \nabla T\right) + S_E \tag{7}$$

$$E = \frac{\sum_{q=1}^{N} \alpha_q \rho_q E_q}{\sum_{q=1}^{N} \alpha_q \rho_q} \tag{8}$$

where E_q—function of the specific heat capacity and temperature of the phase, k_{eff}—effective heat transfer coefficient.

The Nusselt number represents a standard number of the intensity of convection heat transfer, and also represents the ratio of heat conduction resistance at the bottom of fluid laminar flow to convection heat transfer resistance. The friction coefficient f represents the resistance of fluid in the flow channel. The calculation formula [32] is as follows:

$$Nu_x = \frac{hl}{\lambda} \tag{9}$$

$$f = \frac{2 \times \Delta p \times D_e}{L \times \rho \times v^2} \tag{10}$$

The modeled transport equations for κ and ε in the realizable κ-ε model are:

$$\frac{\partial}{\partial t}(\rho k) + \frac{\partial}{\partial x_j}\left(\rho k u_j\right) = \frac{\partial}{\partial x_j}\left[\left(u + \frac{u_t}{\sigma_k}\right)\frac{\partial k}{\partial x_j}\right] + G_k + G_b - \rho\varepsilon - Y_M + S_k \tag{11}$$

$$\begin{aligned} \frac{\partial}{\partial t}(\rho\varepsilon) + \frac{\partial}{\partial x_j}\left(\rho\varepsilon u_j\right) &= \frac{\partial}{\partial x_j}\left[\left(u + \frac{u_t}{\sigma_\varepsilon}\right)\frac{\partial\varepsilon}{\partial x_j}\right] + \rho C_1 S_{\varepsilon b} \\ &\quad - \rho C_2 \frac{\varepsilon^2}{k + \sqrt{v\varepsilon}} + C_{1\varepsilon}\frac{\varepsilon}{k}C_{3\varepsilon}G_b + S_\varepsilon \end{aligned} \tag{12}$$

where $C_1 = \max\left[0.43, \frac{\eta}{\eta+5}\right]$, $\eta = S\frac{k}{\varepsilon}$, $S = \sqrt{2S_jS_j}$.

In these equations, G_k represents the generation of turbulence kinetic energy due to the mean velocity gradients. G_b is the generation of turbulence kinetic energy due to buoyancy. Y_M represents the contribution of the fluctuating dilatation in compressible turbulence to the overall dissipation rate. C_2 and $C_{1\varepsilon}$ are constants. σ_ε and σ_k are the turbulent Prandtl numbers for κ and ε, respectively. S_k and S_ε are user-defined source terms.

2.3. Boundary Conditions and Calculation Settings

Set the upper flow channel as cold fluid and the lower flow channel as hot fluid. Adopt speed inlet and pressure outlet. The cold fluid is water, density 998.2 kg/m^3, viscosity 0.001 kg/m·s, specific heat capacity 4182 J/kg·k, thermal conductivity 0.6 W/m·k. The hot fluid is liquid sulfur, density 2000 kg/m^3, specific heat capacity 23.525 J/kg·k, thermal conductivity 0.269 W/m·k. The velocity of working fluids is 0.01 m/s. The temperature of hot fluid is 450 K, the inlet state is liquid phase, no phase transition occurs. The temperature of the cold fluid is 350 K, the inlet is liquid phase, the second phase is gas phase, and the phase transition occurs when the temperature is 373.15 K. The middle wall is the heat exchange wall, and the uppermost and lower most walls is set as adiabatic walls. The initial fluid flow parameters: Prandtl number = 2.55, Reynolds number = 5000, the hydraulic diameter = 5mm, the turbulence intensities = 2.84525%.

ANSYS software was used to calculate the heat exchanger model. Pressure base solver, transient time, VOF multiphase flow model, Realizeable κ-ε turbulence model were selected, the realizable k-epsilon model has better performance than the standard k-epsilon model in strong streamline bending, vortex and rotation, and is more suitable for flow simulation in plate heat exchanger. The second order upwind scheme is adopted to discretize each governing equation. The relative residual control of the energy governing equation is less than 10^{-6}, and the other governing equations converges on 10^{-5}.

2.4. Model Validation

In order to verify the reliability of the model, numerical simulation of Nu and friction coefficient f was carried out in this paper, which was compared with the experiment of Djordjevic et al. [33]. A unified model of the structure used in the literature experiment was established, and the simulation was carried out with the same inlet Reynolds number. The comparison results between simulation and experiment are shown in Figure 4.

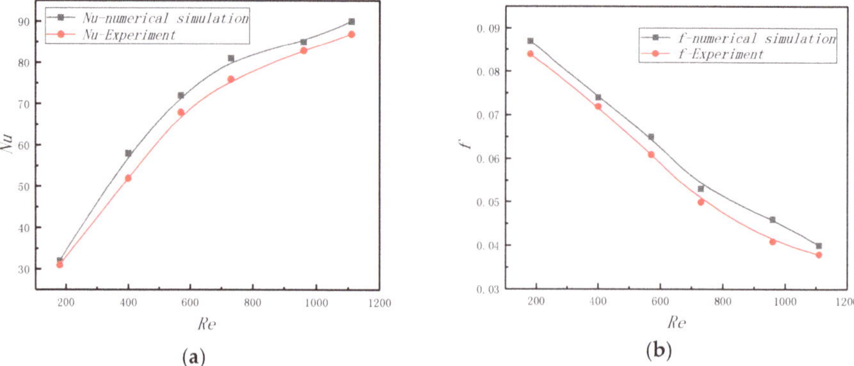

Figure 4. Comparison of simulation result with experiment data for average Nusselt number and friction factor.(**a**) The curve of the Nusselt number (**b**) The curve of the friction factor

Figure 4a is the comparison between *Nu* simulation results and experimental results, and Figure 4b is the comparison between *f* simulation results and experimental results. There is a certain deviation between the simulation and the experiment, which mainly comes from the different dimension of the numerical simulation and the wall surface is more ideal. Compared with the experiment, the deviation of the numerical simulation *Nu* and *f* is less than 5%, which is within the allowable range and the model is reliable.

3. Analysis of Calculation Results

From the time when the fluid starts to flow until the fluid reaches a steady state, the phase transition gradually develops and changes with the increase of time. Therefore, the model of double channel corrugated heat transfer channel was established and the transient simulation was carried out. The time of a fluid from initial to steady state takes T, the steady (the physical process reaches steady state) is when the average temperature of hot and cold fluid outlets no longer changes. The change of phase transition, flow and heat transfer parameters with time in T process was studied.

3.1. Distribution of Fluid Temperature and Phase Volume Distribution

Figure 5 is the temperature distribution diagram in the flow passage at different times. As can be seen from Figure 5, at the initial state of *t* = 0, hot and cold fluid filled the upper and lower flow channels simultaneously, the heat hasn't been transferred yet. The heat of the hot fluid is transferred to the cold fluid through the heat exchange wall. The temperature gradient in the concave corner of the lower wall of the cold fluid passage is denser than that in other regions. As time goes by, the hot fluid flows, and the heat is simultaneously transferred through the wall to the cold fluid. Cold fluid from the inlet to the outlet, the whole process temperature is rising, hot fluid vice versa. The temperature of hot and cold fluid varies in gradient along the channel direction. When *t* = 2/3T, the heat transfer is close to steady, and the temperature change in each region gradually becomes steady. The temperature in the wavy region of the second section of the cold fluid channel first enters the equilibrium state. Throughout the heat transfer process, the cold fluid flowing in first absorbing heat from the hot fluid and transfers it to the cold fluid flowing in later.

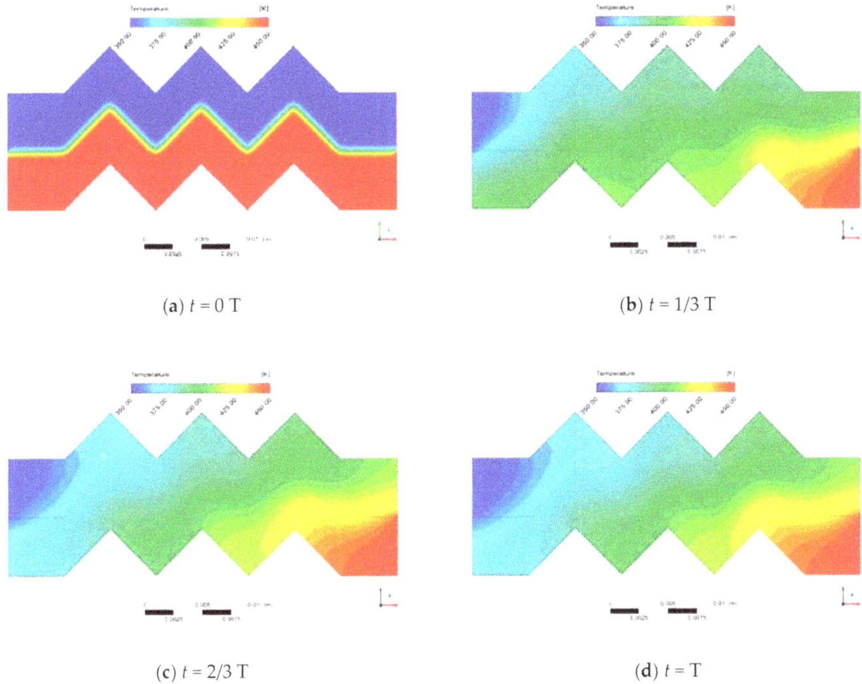

(**a**) $t = 0$ T

(**b**) $t = 1/3$ T

(**c**) $t = 2/3$ T

(**d**) $t = $ T

Figure 5. Temperature distribution at different time.

Figure 6 shows the distribution of gas phase volume fraction in the flow passage of different moments. Figure 6a is the initial time of phase transition. Since the temperature was first transferred to the vicinity of the lower wall surface of the cold flow passage, the gas phase was first generated on the lower wall surface. As can be seen from the figure, the gas phase only exists on the first half of each ripple on the lower wall surface, mainly because the fluid flow process is obstructed here, causing disorder, and the phase change is easy to occur and gather. At the same time, there is a vortex at the concave angle of the ripple, which is more likely to form a gasified core, and the gas phase will also be trapped and collected here. As time grows, the temperature is transferred to the upper wall, the phase transition occurs when the temperature of the fluid flowing near the upper wall reaches the critical temperature.

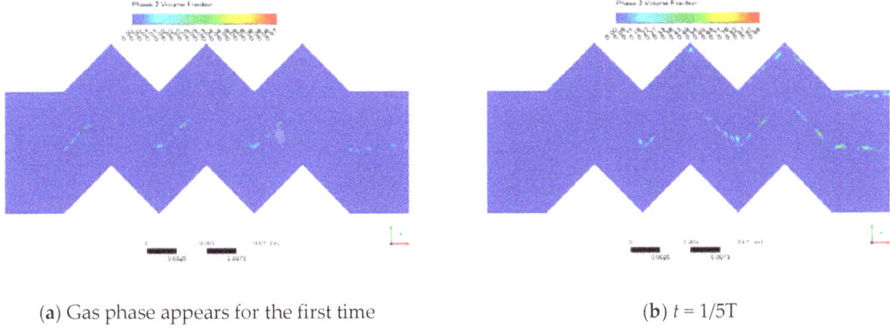

(**a**) Gas phase appears for the first time

(**b**) $t = 1/5$T

Figure 6. *Cont.*

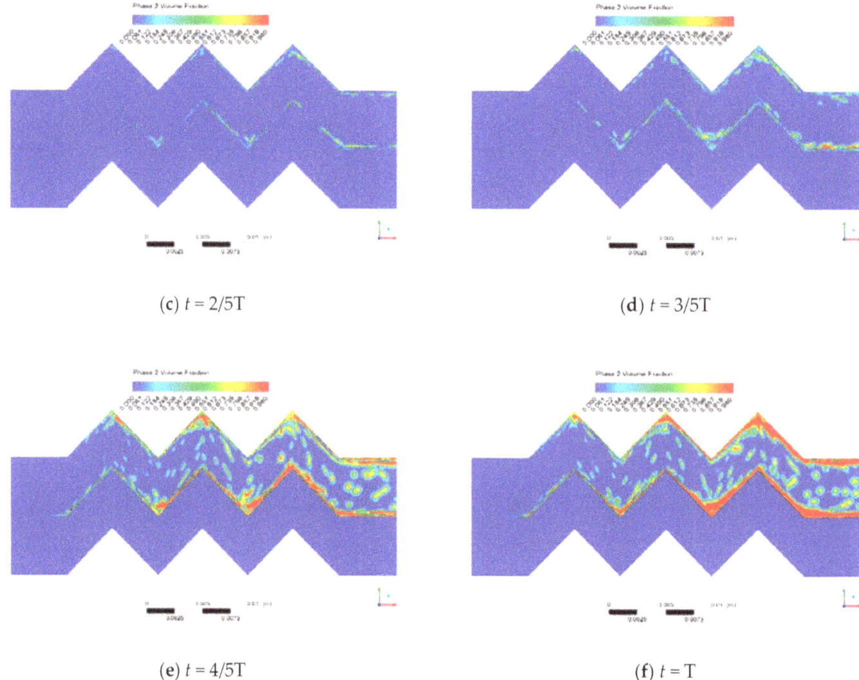

(**c**) $t = 2/5T$ (**d**) $t = 3/5T$

(**e**) $t = 4/5T$ (**f**) $t = T$

Figure 6. Gas phase volume fraction distribution at different time.

It can be seen that the gasified core is formed near the wall surface during the boiling flow. The gas phase in the channel moves slowly towards the exit and gathers along the wall, and the volume fraction of the gas phase increases. The gas phase microparticles at the convex corner were significantly more than those on the straight wall, and the volume fraction was higher than that in other regions.

Figure 7 shows the distribution of velocity vectors in the flow passage of different moments. As can be seen from the figure, when $t = 1/4T$, severe eddies occur in a small range in the flow passage, and the flow velocity in the main flow area is low. When $t = 1/2T$, the range of violent eddies in the flow passage becomes larger and the velocity of eddies becomes larger, and the eddies gradually move to the outlet. Moreover, the flow velocity in the main flow area is significantly higher than when $t = 1/2T$. When $t = 3/4TT$, the range of violent eddies on the upper side of the channel becomes smaller, and the center moves to the near exit. At this point, the flow velocity in the main flow area basically reaches stability. When $t = T$, there is no violent eddy on the upper side of the channel, and the whole fusion becomes a large range of low-velocity reflux.

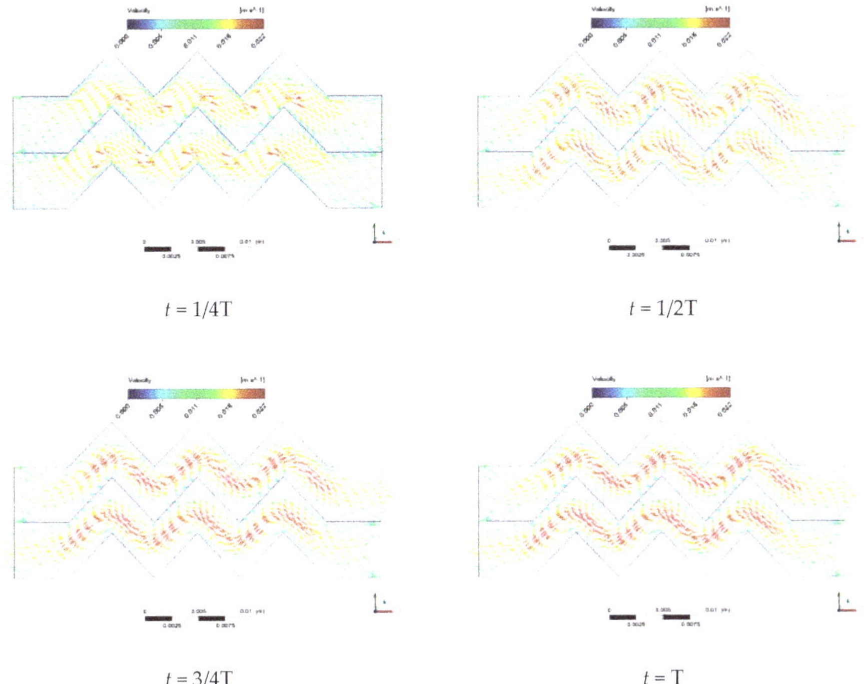

t = 1/4T　　　　　　　　　*t* = 1/2T

t = 3/4T　　　　　　　　　*t* = T

Figure 7. Velocity vectors distribution at different time.

3.2. The Change Rule of Fluid Mean Value Parameters

Figure 8 shows the curve of $\overline{Nu}$ and $\overline{\phi}$ on the heat exchange wall with time. As can be seen from the figure, the change trend of $\overline{Nu}$ and $\overline{\phi}$ on the heat exchange wall is basically consistent. In the process of time t = 1/5T–3/5T, as the temperature difference between hot and cold fluid decreases with the increase of time and the heat transfer decreases, the values of these two average parameters continue to decrease. There is a trough between t = 3/5T–4/5T, because the temperature difference before the trough is large and the heat exchange is intense. When the time trough occurs, the gas phase slowly increases and accumulates to a certain amount, and the latent heat of phase transition begins to affect heat transfer. When t = 4/5T, $\overline{Nu}$ and $\overline{\phi}$ begin to become steady, and the temperature difference fluctuations of cold and hot fluids on both sides of the heat exchange wall gradually decrease, and the heat transfer during phase change also enters a steady state. At the beginning, the heat exchange is more intense, the bubbles are less generated, the heat transfer coefficient is larger, and the Nusselt number and the heat flux are larger. Then the bubble slowly increases, forming a vapor film on the wall to form a thermal resistance. As time went on, more and more bubbles are generated along the wall, the vapor film is thickened, the thermal resistance is increased, the convective heat transfer coefficient decreases, and the Nusselt number and heat flux are reduced. When the heat transfer reaches a stable stage, the vapor film thickness no longer increases, the thermal resistance no longer changes, and the Nusselt number and heat flux do not change.

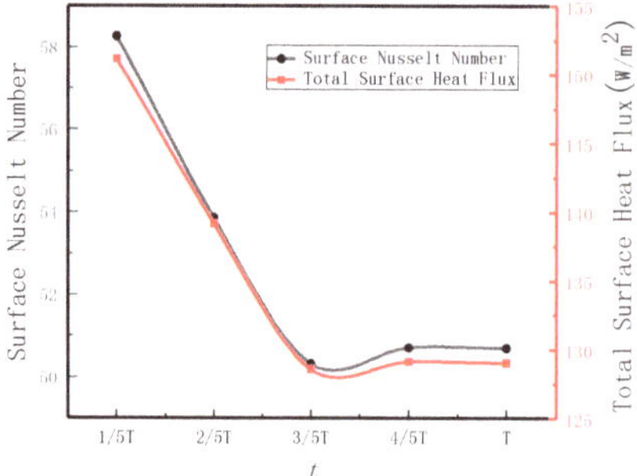

Figure 8. The curve of $\overline{Nu}$ and $\overline{\phi}$ on the heat exchange wall with time

Figure 9 shows the curve of the average temperature at the outlet of the hot and cold fluid with time and the curve of the maximum volume fraction of wet saturated steam at the outlet of the phase change passage with time. Outlet 1 is the cold fluid outlet, and outlet 2 is the hot fluid outlet. The average temperature of cold fluid outlet is 406 K–420 K, and the average temperature of hot fluid outlet is 366 K–390 K. In the process of time t = 1/5T–3/5T, the average temperature of the cold fluid outlet rose rapidly, the average temperature of the hot fluid outlet decreased rapidly, and the volume fraction of the gas phase only increased slightly. After t = 3/5T, the average temperature of hot and cold fluid outlet began to stabilize, while the volume fraction of the gas phase began to increase sharply. At this time, the temperature of most fluid areas in the cold passage is higher than the phase transition temperature, and the liquid along the wall surface is subjected to violent boiling and vaporization. The number of gas phases attached to the wall surface increased, and the number of gas phases converging to the outlet, which overcame the wall tension, also increased. Finally, the volume fraction of the gas phase was 1, and the vicinity of the outlet wall surface was all gas phase. When the hot and cold fluid is steady, the average temperature difference at the two outlets is similar to the average temperature difference at the initial state. Figures 7 and 8 mutually demonstrate that the process of heat transfer during phase change in PHE tends to a steady state after t = 3/5T.

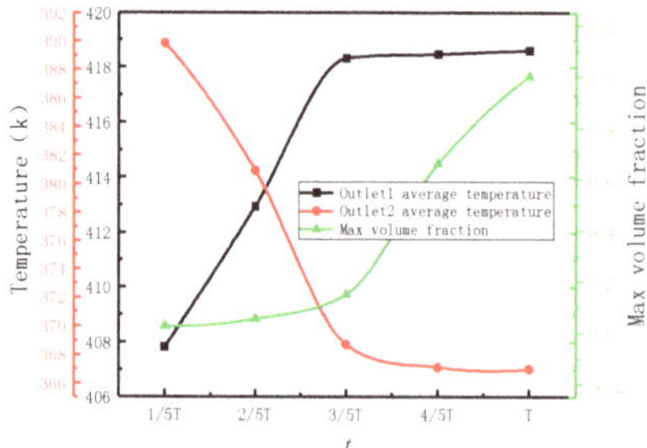

Figure 9. The curve of temperature at the outlet of hot and cold fluid and the vapor maximum volume fraction of wet saturated steam at the outlet changing with time.

3.3. Fluid Outlet Temperature Change Rule

In order to verify the direction of heat transfer, the temperature distribution change of fluid outlet was investigated. Figure 10 shows the temperature distribution of the cold fluid outlet along the Y direction at different moments. Figure 11 shows the temperature distribution of the hot fluid outlet along the y-direction at different moments. At the beginning of fluid flow, the heat transfer between hot and cold fluid is larger, resulting in a larger temperature gradient at the outlet. The temperature at one end of the cold fluid outlet in contact with the heat exchange wall is higher than that at the other end of the non-heat exchange wall. As time grows, the overall temperature at the y-direction of the cold fluid outlet is rising, while the overall temperature at the Y direction of the hot fluid outlet is decreasing, and the temperature gradient of both is decreasing. The temperature difference between the heat exchange wall and the non-heat exchange wall is decreasing. Throughout the flow, in the Y direction of the channel, heat is transferred from the bottom up, from the high temperature fluid field to the low temperature fluid field.

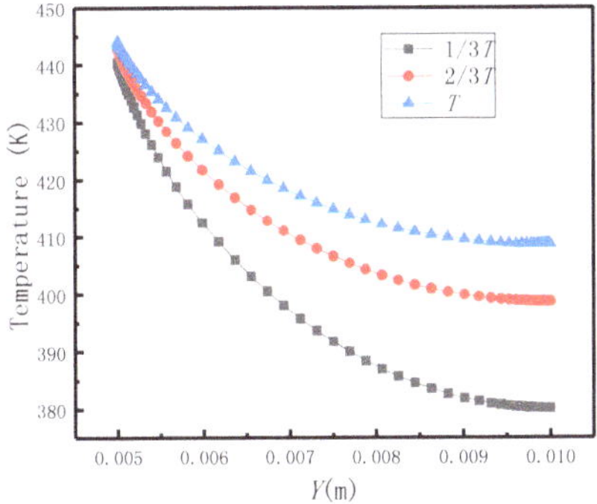

Figure 10. Temperature distribution along the outlet Y direction of cold fluid at different time.

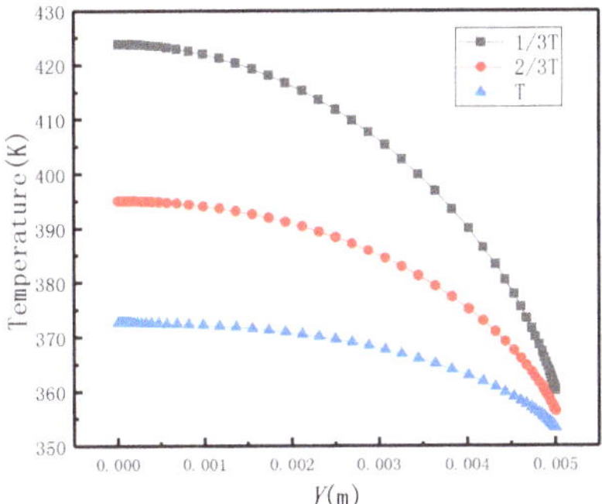

Figure 11. Temperature distribution along the outlet Y direction of thermal fluid at different time.

3.4. The Change Rule of Heat Exchange Wall Surface Parameters Along the Path

From the change of mean value parameters and fluid outlet temperature, the time to reach stability in PHE and the direction of heat transfers can be obtained, but the state of heat exchanges wall surface between cold and hot fluids cannot be reflected. Figure 12 shows the temperature distribution along the heat transfer wall of corrugated plate at different moments. As can be seen from the figure, the temperature of the heat exchange wall fluctuates and rises along the X-axis in the whole process, and the temperature fluctuation corresponds to the turning point of the corrugated plate. The temperature crest is at the concave angle of the corrugated plate structure, and the trough is at the convex angle of the corrugated plate. When $t = 1/3$ T, the temperature distribution of the second and third sections of the corrugated plate is consistent. Since the vortex first occurs to the concave angle between the ripples in the second and third sections, the temperature boundary layer thins and the heat transfer in the third section becomes faster. As time grows, the overall temperature of the corrugated heat transfer wall decreases slightly. The temperature at the back end reached a steady state before that at the front end, and the temperature gradient at the inlet and outlet end was significantly larger than that at the middle end, which was due to the intense heat transfer of the inlet and outlet fluid and the great change of the wall surface temperature.

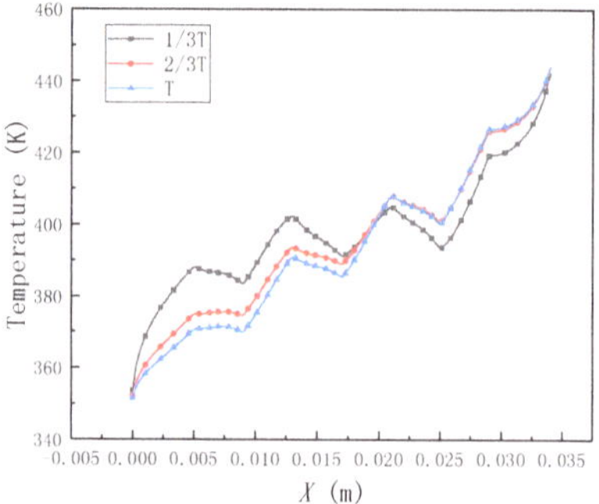

Figure 12. Temperature distribution along the heat transfer wall at different times.

Figure 13 shows *Nu* distribution along the heat transfer wall of the middle corrugated plate at different times. As can be seen from the figure, the distribution of *Nu* along the heat exchange wall surface to fluctuate on the whole, and fluctuates violently around the corrugated corner. At different times, the change trend of *Nu* distribution curve is basically the same, with small local differences, and the change rate of curve is basically the same. The curve of the inlet section of the cold fluid drops steeply from the extreme value, fluctuates gently along the X-axis, and rises steeply to the extreme value at the outlet, which is greater than the peak value of the fluctuation section, but less than the maximum value of the inlet. At the entrance to the cold fluid, the entry into the cold fluid leads to the destruction of the boundary layer, where the heat is strongly mixed and diffused, and the heat exchange intensity suddenly increases. At the outlet of the cold fluid, the cold fluid flows out and the hot fluid flows in, the temperature gradient increases sharply, and the heat transfer coefficient reaches the maximum value here. As time grows, the extreme value of Nu at the inlet and outlet decreases, because the temperature difference between hot and cold fluid in the flow channel decreases, which weakens the previous effect.

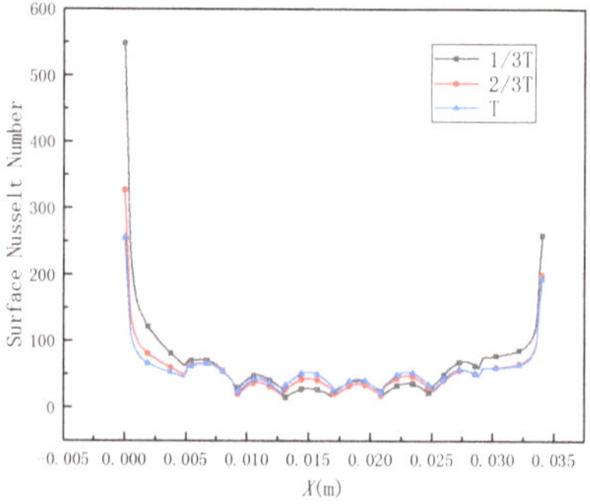

Figure 13. *Nu* distribution along the heat transfer wall at different times.

Figure 14 shows the φ distribution diagram along the heat exchange wall of the middle corrugated plate at different times. As can be seen from the figure, the change trend of φ distribution along the wall surface is close to that of *Nu* distribution curve, and the overall change is fluctuant, which fluctuates sharply near the corrugated corner. The curve of the inlet sections drops steeply from the maximum value, which is less than the extreme value of the outlet, because the heat flux is related to the temperature difference, and the temperature difference between the cold and hot fluid at the inlet wall is less than that at the outlet wall. At different times, the change trend of the heat flux distribution curve is basically the same, and the fluctuation distribution in the middle section increases slightly with the increase in time, because the temperature difference in both sides with the heat exchange wall is decreasing.

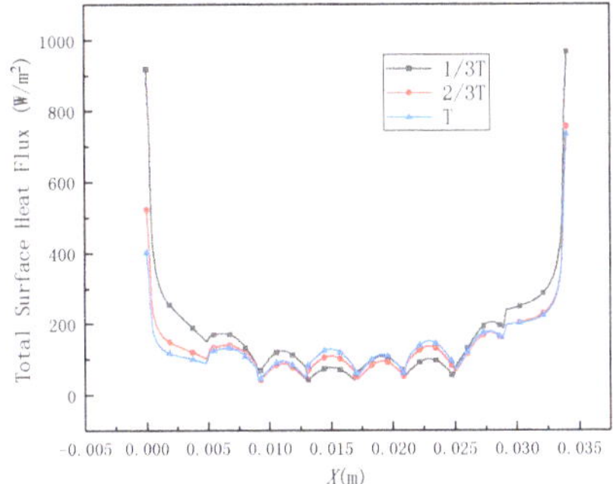

Figure 14. Heat flux distribution along the heat transfer wall at different times.

4. Conclusions

In this paper, the process of heat transfer during phase change in corrugated plate heat exchanger flow passage, the change law of heat transfer during phase change average parameters and boundary local parameters at different moments are studied, and the following conclusions are drawn:

(1) The $\overline{Nu}$ and $\overline{\phi}$ of the heat exchange wall surface decrease with time, reaching stability at $t = 3/5T$. The average temperature of the cold fluid outlet increased with time, while the average temperature of the hot fluid outlet was opposite, but both reached stability when $t = 3/5T$. For the volume fraction of the gas phase, it was small before $t = 3/5T$, and then it increased dramatically.

(2) In the cold and hot fluid passages, the temperature gradient distribution is thin in the middle and dense at both ends. Heat is transferred from the bottom to up, from high temperature to low temperature.

(3) For the heat exchange wall, the temperature fluctuates along the X-axis. The peak value is at the concave angle of ripple, and the wall surface temperature in the later half tends to be steady firstly. For the cold fluid, heat transfer is faster at the concave corner of the heat exchange wall surface, and vaporization core is easier to form.

(4) Compared with the cold fluid, the temperature distribution of the hot fluid near the inlet reaches stability first. The change of heat exchange wall surface along Nu and φ is basically the same, reaching the maximum at the inlet and outlet. The middle section fluctuates along the direction of the ripples, and its value increases slightly as time.

(5) The unsteady calculation model was established, and the gas-liquid phase change process between fluids was realized by using the heat and mass transfer model. The evolution law of heat transfer during phase change was found. The method is innovative. Further study on the influencing factors of heat transfer during phase change is needed in the future.

Author Contributions: Conceptualization, H.S.; Data curation, C.H.; Formal analysis, W.Y. and Z.G.; Methodology, G.W.; Software, T.H.; Writing—original draft, H.S. and C.H.; Writing—review & editing, C.H. All authors have read and agreed to the published version of the manuscript.

Funding: This research was funded by [Science and Technology Research Project under the Educational Commission of Hubei Province of China] grant number [Q20191208].

Conflicts of Interest: The authors declare no conflict of interest.

Data Availability: The numerical data used to support the findings of this study are included within the article.

Nomenclature

λ	The spacing of the ripples, mm
H	The height of the ripples, mm
β	The Angle of the ripples, (°)
Re	Reynolds number
f	The coefficient of friction
Nu	Nusselt number
$\overline{Nu}$	Average Nusselt number
t	Time, T
ϕ	Heat flux, W/m^2
$\overline{\phi}$	Average heat flux, W/m^2
α_v	Vapor phase volume fraction
ρ_v	Vapor phase density, kg/m^3
α_l	Liquid volume fraction
ρ_l	Liquid density, kg/m^3
u	Vapor phase velocity, m/s
T_l	Liquid phase temperature, K
T_{sat}	Phase transition temperature, K
T_v	Vapor phase temperature, K
β	Relaxation factor
S_M	The quality of the source term
ΔH	Enthalpy
ρ	Density, kg/m^3
μ	Dynamic viscosity, m^2/s
Eq	Function of the specific heat capacity and temperature of the phase
$keff$	Effective thermal conductivity
S_E	Energy source term

References

1. Eldeeb, R.; Vikrant, A.; Reinhard, R. A survey of correlations for heat transfer and pressure drop for evaporation and condensation in plate heat exchangers. *Int. J. Refrig.* **2016**, *5*, 12–26. [CrossRef]
2. Mcdonald, A.-G.; Magande, H.-L. Fundamentals of heat exchanger design. In *Introduction to Thermo-Fluids Systems Design*; John Wiley & Sons, Ltd.: Hoboken, NJ, USA, 2012.
3. Mota, F.A.; Ravagnani, M.A.; Carvalho, E.P. Optimal design of plate heat exchangers. *Appl. Therm. Eng.* **2014**, *63*, 33–39. [CrossRef]
4. Qiao, H.; Aute, V.; Lee, H.; Saleh, K.; Radermacher, R. A new model for plate heat exchangers with generalized flow configurations and phase change. *Int. J. Refrig.* **2013**, *36*, 622–632. [CrossRef]
5. Kumar, V.; Tiwari, A.-K.; Ghosh, S.-K. Application of nanofluids in plate heat exchanger: A review. *Energy Convers. Manag.* **2015**, *105*, 1017–1036. [CrossRef]
6. Gulenoglu, C.; Akturk, F.; Aradag, S.; Uzol, N.S.; Kakac, S. Experimental comparison of performances of three different plates for gasketed plate heat exchangers. *Int. J. Therm. Sci.* **2014**, *75*, 249–256. [CrossRef]
7. Faizal, M.; Ahmed, M.-R. Experimental studies on a corrugated plate heat exchanger for small temperature difference applications. *Exp. Therm. Fluid Sci.* **2012**, *36*, 242–248. [CrossRef]
8. Lee, J.; Lee, K.-S. Flow characteristics and thermal performance in chevron type plate heat exchangers. *Int. J. Heat Mass Transf.* **2014**, *78*, 699–706. [CrossRef]
9. Tiwari, A.-K.; Ghosh, P.; Sarkar, J. Heat transfer and pressure drop characteristics of CeO$_2$/water nanofluid in plate heat exchanger. *Appl. Therm. Eng.* **2013**, *57*, 24–32. [CrossRef]
10. Goodarzi, M.; Amiri, A.; Goodarzi, M.-S.; Safaei, M.R.; Karimipour, A.; Languri, E.M.; Dahari, M. Investigation of heat transfer and pressure drop of a counter flow corrugated plate heat exchanger using MWCNT based nanofluids. *Int. Commun. Heat Mass Transf.* **2015**, *66*, 172–179. [CrossRef]
11. Tiwari, A.K.; Ghosh, P.; Sarkar, J.; Dahiya, H.; Parekh, J. Numerical investigation of heat transfer and fluid flow in plate heat exchanger using nanofluids. *Int. J. Therm. Sci.* **2014**, *85*, 93–103. [CrossRef]

12. Demir, H.; Dalkilic, A.-S.; Kürekci, N.-A.; Duangthongsuk, W.; Wongwises, S. Numerical investigation on the single phase forced convection heat transfer characteristics of TiO_2 nanofluids in a double-tube counter flow heat exchanger. *Int. Commun. Heat Mass Transf.* **2011**, *38*, 218–228. [CrossRef]

13. Guo, Z.J.; Hu, C.Q.; Liu, J.R.; Su, H.S. The Study on Internal Flow Characteristics of Magnetic Drive Pump. *IEEE Access* **2019**, *7*, 100003–100013.

14. Yildiz, A.; Mustafa, A.E. Theoretical and experimental thermodynamic analyses of a chevron type heat exchanger. *Renew. Sustain. Energy Rev.* **2015**, *42*, 240–253. [CrossRef]

15. Grabenstein, V.; Polzin, A.E.; Kabelac, S. Experimental investigation of the flow pattern, pressure drop and void fraction of two-phase flow in the corrugated gap of a plate heat exchanger. *Int. J. Multiph. Flow* **2017**, *91*, 155–169. [CrossRef]

16. Manglik, R.-M. Plate heat exchangers: Design, applications and performance. *Mech. Eng.* **2007**, *5*, 58.

17. Khan, T.S.; Khan, M.S.; Chyu, M.C.; Ayub, Z.H.; Chattha, J.A. Review of heat transfer and pressure drop correlations for evaporation of fluid flow in plate heat exchangers (RP-1352). *HVACR Res.* **2009**, *15*, 169–188. [CrossRef]

18. Li, M.C.; Xiao, G.; Shi, H.C.; Liu, H.-L.; Chen, D.; Luo, Z.-Y. Study on phase change flow and heat transfer characteristics of chevron corrugated plate. *J. Eng. Thermophys.* **2016**, *37*, 581–585.

19. Huang, J.; Sheer, T.-J.; Bailey-Mcewan, M. Heat transfer and pressure drop in plate heat exchanger refrigerant evaporators. *Int. J. Refrig.* **2011**, *34*, 325–335. [CrossRef]

20. Longo, G.A.; Righetti, G.; Zilio, C. A new computational procedure for refrigerant condensation inside herringbone-type Brazed Plate Heat Exchangers. *Int. J. Heat Mass Transf.* **2015**, *82*, 530–536. [CrossRef]

21. Longo, G.A.; Zilio, C. Condensation of the low GWP refrigerant HFC1234yf inside a brazed plate heat exchanger. *Int. J. Refrig.* **2013**, *36*, 612–621. [CrossRef]

22. Longo, G.-A.; Zilio, C.; Righetti, G.; Brown, J.S. Experimental assessment of the low GWP refrigerant HFO-1234ze(Z) for high temperature heat pumps. *Exp. Therm. Fluid Sci.* **2014**, *57*, 293–300. [CrossRef]

23. Longo, G.A.; Mancin, S.; Righetti, G.; Zilio, C. A new model for refrigerant boiling inside Brazed Plate Heat Exchangers (BPHEs). *Int. J. Heat Mass Transf.* **2015**, *91*, 144–149. [CrossRef]

24. Longo, G.A.; Mancin, S.; Righetti, G.; Zilio, C. HFC404A condensation inside a small brazed plate heat exchanger: Comparison with the low GWP substitutes propane and propylene. *Int. J. Refrig.* **2017**, *81*, 41–49. [CrossRef]

25. Han, D.-H.; Lee, K.-J.; Kim, Y.-H. Experiments on the characteristics of evaporation of R410A in brazed plate heat exchangers with different geometric configurations. *Appl. Therm. Eng.* **2003**, *23*, 1209–1225. [CrossRef]

26. Mancin, S.; Del-Col, D.; Rossetto, L. Condensation of superheated vapour of R410A and R407C inside plate heat exchangers: Experimental results and simulation procedure. *Int. J. Refrig.* **2012**, *35*, 2003–2013. [CrossRef]

27. Mancin, S.; Del Col, D.; Rossetto, L. R32 partial condensation inside a brazed plate heat exchanger. *Int. J. Refrig.* **2013**, *36*, 601–611. [CrossRef]

28. Wu, X.-H.; Li, C.; Gong, Y.; Zhang, J.; Zhao, M. Heat transfer and pressure drop characteristics of phase change flow in plate heat exchanger. *Chin. J. Chem. Eng.* **2017**, *68*, 133–140.

29. Yu, J.Z. *Principle and Design of Heat Exchanger*; Beijing University of Aeronautics and Astronautics Press: Beijing, China, 2006.

30. Wang, F.J. *Analysis Computational Fluid Dynamics: Principles and Applications of CFD Software*; Tsinghua University Press: Beijing, China, 2004.

31. DE Schepper, S.C.; Heynderickx, G.J.; Marin, G.B. CFD modeling of all gas-liquid and vapor-liquid flow regimes predicted by the Baker chart. *Chem. Eng. J.* **2008**, *138*, 349–357. [CrossRef]

32. Derome, D.; Blocken, B.; Carmeliet, J. Determination of surface convective heat transfer coefficients by CFD. In Proceedings of the 11th Canadian Conference on Building Science and Technology, Banff, AB, Canada, 22–23 March 2007.

33. Djordjevic, E.; Kabelac, S. Flow boiling of R134a and ammonia in a plate heat exchanger. *Int. J. Heat Mass Transf.* **2008**, *51*, 6235–6242. [CrossRef]

Article

Effect of Orientation and Aspect Ratio of an Internal Flat Plate on Natural Convection in a Circular Enclosure

Anjie Wang [1,2], Cunlie Ying [3], Yingdong Wang [3], Lijun Yang [3], Yunjian Ying [3], Lulu Zhai [1,2] and Wei Zhang [1,2,*]

[1] National-Provincial Joint Engineering Laboratory for Fluid Transmission System Technology, Zhejiang Sci-Tech University, Hangzhou 310018, Zhejiang, China; wang_st154@163.com (A.W.); zhailulu@zstu.edu.cn (L.Z.)

[2] Faculty of Mechanical Engineering and Automation, Zhejiang Sci-Tech University, Hangzhou 310018, Zhejiang, China

[3] Zhejiang Yilida Ventilator Co., Ltd., Taizhou 318056, Zhejiang, China; yingcunlie@yilida.com (C.Y.); wangyingdong@yilida.com (Y.W.); yanglijun@yilida.com (L.Y.); yingyunj@163.com (Y.Y.)

* Correspondence: zhangwei@zstu.edu.cn

Received: 8 November 2019; Accepted: 25 November 2019; Published: 2 December 2019

Abstract: This work presents a numerical investigation on natural convection in a circular enclosure with an internal flat plate at $Ra = 10^6$. The cross-section area of the plate was fixed at three values, $H{\cdot}W/D^2 = 0.01, 0.04$, and 0.09, in which H and W are the height and width of the plate and D is the diameter of the enclosure while the aspect ratio changes, which makes the plate vertically placed ($H > W$) or horizontally placed ($H < W$). The objective of this work was to explore the effects of the orientation and aspect ratio of the plate on the characteristics of natural convection in various aspects. The numerical results reveal that the overall heat transfer rate is higher for the vertically placed plate and increases with the cross-section area, while the width of the plate has almost no effect for the horizontally placed plate, especially for the plate with a relatively large cross-section area. Depending on the orientation and aspect ratio, there can be one primary vortex, one primary and one secondary vortex, or one secondary and two separated vortices to each side of the plate, and the thermal plume structure may appear at the sharp top corners of the plate. Consequently, local heat transfer on the surfaces of the enclosure and plate is affected. Synergy analysis reveals that the enhancement of heat transfer from the fluid circulation is the most significant at the center of the vortices and at the boundary between them.

Keywords: natural convection; flat plate; aspect ratio; orientation; vertical; horizontal

1. Introduction

Natural convection in an enclosure is a fundamental problem in many engineering applications. For example, in sand casting of a metal component, the liquid metal experiences cooling from the sand mold and weakly circulates before its solidification; for indoor air-conditioning in the winter, heaters induce fluid circulation and alter the temperature distribution in a room. In these applications, the buoyancy that drives the fluid's circulation within the enclosed space is generated by the temperature difference between either the various sidewalls of the enclosure (differentially heated enclosure) or the enclosure and the internal entity. For the latter case, the characteristics of fluid flow and heat transfer are determined by a number of factors, including the shape, size, and position of the internal entity within the enclosure; the temperature difference between the entity and the enclosure; the thermal boundary conditions, and the thermophysical properties of the fluid. The thermal and flow processes are integrated, in that the flow is induced by the temperature difference, while the circulating flow results in mass transfer and also influences the temperature distribution. Although there have been many studies on this topic, the problem is still worthy of investigation for specific geometrical configurations considering the complex and various configurations encountered in realistic applications.

The natural convection between a cold enclosure and a hot internal thin flat plate is a model as an approximation to realistic applications. The enclosure is normally assumed have circular or rectangular geometry for production simplicity, while the internal plate can be placed by attaching to the sidewalls of the enclosure or in an isolated way. Depending on the position and orientation of the plate, the pattern of fluid circulation is different, and the heat transfer is consequently affected. Alteç and Kurtul [1] studied natural convection in a tilted rectangular enclosure with an internal isolated flat plate at $Ra = 10^5$–10^7. The enclosure has three adiabatic sidewalls and one cold sidewall, which was always parallel with the plate. It was found that for a square enclosure at a high Rayleigh number, the mean heat transfer rate increased with the tilt angle and reached a maximum at 22.5°; however, the mean heat transfer rate remained almost unchanged up to 22.5° for the enclosure with an aspect ratio of two. A similar physical configuration was also studied by Wang et al. [2]. The tilted enclosure has two adiabatic sidewalls, and two cold sidewalls, which can either be parallel with or perpendicular to the internal hot flat plate. Heat transfer is enhanced by the vertically placed plate. The effect of the two orientations of the internal plate was also studied by Öztop et al. [3] for a horizontally placed enclosure with cold left and right sidewalls at $Ra = 10^4$–10^6 by considering various sizes and positions of the plate. The mean heat transfer rate monotonically increased with the size of the plate and the increase was more notable for the vertically placed plate at all Rayleigh numbers. Tasnim and Collins [4] considered natural convection in a differentially heated square enclosure with an internal adiabatic arc-shaped plate. The length and radius of the arc were varied to explore the blockage effect on the flow and thermal behaviors. It was found that the arc modified the pattern of fluid circulation, and the modification was enhanced at small arc radius; the overall heat transfer was degraded by the introduction of the arc. For a finite-thickness rectangular plate placed within a square enclosure, Wang et al. [5] found that the distance between the plate and the sidewall of the enclosure is a critical parameter. For $Ra < 5 \times 10^5$, the overall heat transfer rate is not sensitive to the position of the plate as the distance is larger than about one fourth of the cavity width while it significantly increases with the decreasing distance as the distance is smaller than 20% of the cavity width.

There are also a number of works focusing on natural convection in an enclosure with multiple plates. The fluid circulation is confined by the plates, and the formation and interaction of recirculating vortices are complexly dependent on the geometrical parameters and thermal properties of the plates. Hakeem et al. [6] studied natural convection in a square enclosure with two heat-generating flat plates of the same size but positioned perpendicular to each other, i.e., one was vertically placed and the other was horizontally placed. The effect of the location of the plates was numerical studied. It was concluded that the overall heat transfer rate is not significantly affected by the movement of either plate as long as the plates are not wall mounted. In the condition that the enclosure surface was prescribed with a constant heat flux, the heat transfer was degraded by the upward movement of the horizontal plate but enhanced by the horizontal movement of the vertical plate. In the following work, Kandaswamy et al. [7] explored the same physical problem except for thermal boundary conditions in which the enclosure was held at a constant low temperature while the temperature of the two plates was different. They found that the heat transfer mechanism was mainly determined by the hotter plate, and the overall heat transfer rate was higher as the vertical plate was hotter than the horizontal one.

In addition to the dimension and position, the inclination of the internal plate also affects the natural convection in an enclosure. Singh and Liburdy [8] experimentally investigated the natural convection in a circular enclosure with an internal thin flat plate using the holographic interferometry technique. The authors found that the local and overall heat transfer rate is strongly affected by the inclination since the pattern of flow separation at the ends of the plate is greatly influenced. The effect of inclination is minor for angles greater than 60 degrees from the horizontal. Recently, Zhang et al. [9] numerically investigated natural convection in a circular enclosure with an internal flat plate that was inclined and eccentrically placed. The formation and intensity of recirculating vortices are dependent on the eccentricity and inclination, and there can be up to two and three vortices in the left and right halves of the enclosure, respectively. It was found that the heat transfers in the gap between the end of the plate and the enclosure is dominated by conduction at high eccentricity.

The internal plates determine the heat transfer characteristics by changing the thermal conditions but also through modification of the fluid circulation pattern. For isolated plates placed within the enclosure, the fluid circulation is confined but not fully suppressed; however, as the plates are mounted on the walls of the enclosure, recirculating vortices may form in the corner region and greatly alter the local heat transfer performance. Dagtekin and Öztop [10] studied natural convection in an enclosure with two hot plates mounted on the bottom wall and investigated the effect of height and spacing on the heat transfer. The mean heat transfer rate increased with the height of the plate because of the increased surface area of the heating source, and the position of the plates had more of an effect on the flow than on the heat transfer. Nag et al. [11] considered an infinite thermal-conductive or adiabatic plate mounted on the hot sidewall of a differentially heated square enclosure. The authors concluded that the plate of infinite thermal conductivity always increases the overall heat transfer rate irrespective of its position or size while the adiabatic plate attenuates the heat transfer. The same physical configuration was also studied by Tasnim and Collins [12] to investigate the effects of length and position on natural convection. Two competing mechanisms were observed, which determined the patterns of fluid flow and heat transfer, i.e., the blockage effect, which depends on the length of the plate; and the heating of the fluid by the plate. For a short triangular thermal-conductive fin, Sun et al. [13] studied the mixed convection in a lid-driven cavity by mounting the fin on the hot sidewall, cold sidewall, or adiabatic bottom wall. It was found that by placing the fin on the left or right sidewall, the effect of the fin on the heat transfer performance was not only determined by its position but was also greatly dependent on the moving direction of the lid of the cavity through the interaction between the buoyancy generated by the temperature difference and viscous shear stress of the lid. However, the fin on the bottom wall had a tiny effect on the streamline and temperature distributions. The effect of natural convection on the flow in an enclosed space has also been considered in other applications [14–17].

It is summarized from the literatures reviewed above that for a hot plate in a cold enclosure, the natural convection is influenced by a number of factors, including the dimension, position, orientation, and shape of the plate, and the associated thermal boundary conditions for both the plate and enclosure. The existence of the plate affects the fluid flow and heat transfer primarily through two mechanisms. The first is the heating of fluid on the surface, which is the driving force for the fluid circulation; the second is the confinement on the fluid circulation by partially partitioning the enclosure into multiple connected sub-domains, thus the behaviors of the local evolution of vortices are greatly altered. Compared with the cylinder, the geometry of a plate is characterized by its high specific surface area, where the heating of fluid around it is more intense but may not be more effective depending on the pattern of fluid circulation, which is affected by the orientation of the plate and the confinement imposed on the fluid. At medium and high Rayleigh numbers, the local heat transfer performance is significantly lowered by the local quasi-stationary fluid in the corner region or the separated bubble. In most of the existing researches reviewed above, the plate is assumed as a zero-thickness one, which only heats the fluid and confines the flow while the effect of shape and size has not been thoroughly discussed. In this work, we performed a numerical investigation on the natural convection in a circular enclosure with an internal flat plate of various dimensions. Assuming that the plate is produced with a certain amount of material, the cross-section area is fixed but the height and width are varied. The plate is placed within the enclosure in two orientations depending on its height (H, dimension in the vertical direction) and width (W, dimension in the horizontal direction): The vertically positioned plate for $H > W$ and horizontally positioned plate for $H < W$. It is anticipated that by varying the height and width of the plate, the fluid is heated and circulates within the enclosure in various modes, resulting in different distributions of the local heat transfer rate and recirculating vortices. The above flat plate geometry is a typical model widely encountered in engineering applications of enhanced/weakened heat transfer. For example, in the transportation of certain types of liquid in heat pipes in small- and mid-sized electronic devices, the pipe containing the liquid is normally enclosed by an outer enclosure to reduce heat exchange with the surrounding medium. The choice of the geometry of the cross-section of the heat pipe is significant and two factors have to be considered. The first is that the area of the cross-section cannot be reduced to permit a sufficient mass flow rate, thus the velocity of the internal liquid could remain constant, which facilitates the design of the accessory components. The second is that the geometry of the heat pipe is normally a rectangle to save space, which facilitates the layout of multiple heat pipes in a small space. In this condition, the maximum/minimum heat transfer rate between the heat pipe and outer enclosure is a quantity that is determined by the geometry of the heat pipe. The configuration is also employed in other applications. In heating tubes with the thin plate twined with an electric resistance wire as the thermal source, the effectiveness of heating, as measured by the heat transfer rate and uniform temperature rise in the whole enclosed space, is dependent on the position, orientation, and size of the plate. The thermal treatment of the metal component after welding is also characterized by natural convection of the internal entity in an enclosed space. The present work is a theoretical research on a flat plate of various geometries, which has not been studied before. The objective of the present study was to qualitatively and quantitatively investigate the effect of the dimensions and shape of a plate on the thermal and flow characteristics. The effects are presented and analyzed by the overall heat transfer rate, the spatial structures of isotherms and streamlines, and the distributions of the local heat transfer rate on the surfaces of the plate and enclosure. We also carried out synergy principle analysis to visualize how convection enhances heat transfer under different parameter combinations.

2. Numerical Setup

2.1. Physical Model

The configuration of the physical problem is shown in Figure 1. A flat plate was placed concentrically within a circular enclosure. The height of the plate is H and the width is W, and the radius of the circular enclosure is R and diameter is D. Depending on the values of H and W, we considered three types of plate, namely, square ($H = W$), vertically positioned ($H > W$), and horizontally positioned ($H < W$) plate. The three types were assumed to have the same cross-sectional area, i.e., constant cross-section area $H{\cdot}W$, with the consideration that the production of the various plates would cost the same amount of material. The surfaces of the plate and enclosure are isothermal and denoted by the higher temperature, T_i, and lower temperature, T_o, respectively. The enclosure was filled with air ($Pr = 0.71$). All simulations were performed at Rayleigh number $Ra = 10^6$, where heat transfer is dominated by natural convection. In this work, the cross-section area of the plate was chosen at $H{\cdot}W/D^2 = 0.01, 0.04,$ and 0.09. The values for the height and width of the three types of plate are listed in Table 1 with notations clearly explained.

Table 1. Notations for the flat plate of various geometries in the present study. The first letter S, M, and L respectively denotes small, medium, and large sizes; the second letter H, V, and S respectively denotes horizontally positioned, vertically positioned, and square plates.

Notation	$H{\cdot}W/D^2$	Remark
SV	0.01	$H > W$ and $H/D = 0.2\,(0.1)\,0.8$
SH	0.01	$H < W$ and $W/D = 0.2\,(0.1)\,0.8$
SS	0.01	$H/D = W/D = 0.1$
MV	0.04	$H > W$ and $H/D = 0.3\,(0.1)\,0.8$
MH	0.04	$H < W$ and $W/D = 0.3\,(0.1)\,0.8$
MS	0.04	$H/D = W/D = 0.2$
LV	0.09	$H > W$ and $H/D = 0.4\,(0.1)\,0.8$
LH	0.09	$H < W$ and $W/D = 0.4\,(0.1)\,0.8$
LS	0.09	$H/D = W/D = 0.3$

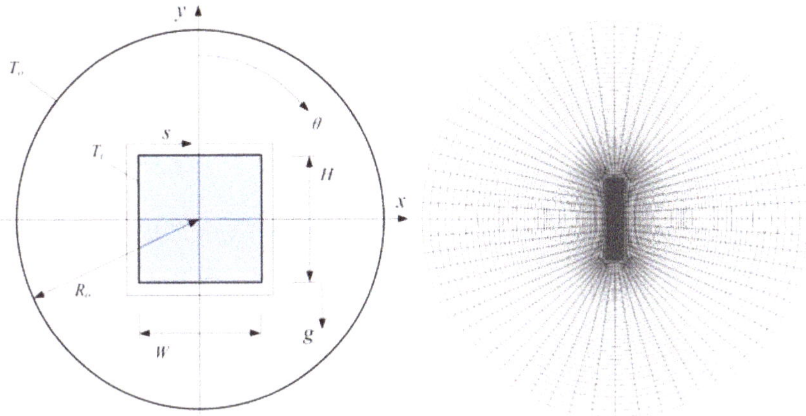

Figure 1. (**Left**) Configuration of the physical problem. The height of the plate is H and the width is W; s is the local coordinate along the cylinder surface. (**Right**) A schematic of the structured grid for the plate at $H/D = 0.2$ and $W/D = 0.05$ (denoted as $SV_{0.2}$, i.e., Small-size and Vertically positioned, in the following figures); the grid is plotted at eight gridlines in the circumferential direction and four gridlines in the radial direction for clarity.

2.2. Numerical Methods

The thermal and flow patterns are governed by the two-dimensional equations of mass, momentum, and energy:

$$\nabla \cdot \mathbf{u} = 0, \tag{1}$$

$$\frac{\partial \mathbf{u}}{\partial t} + (\mathbf{u} \cdot \nabla)\mathbf{u} = -\nabla p + \sqrt{\frac{Pr}{Ra}}\nabla^2 \mathbf{u} + (0, T)^T, \tag{2}$$

$$\frac{\partial T}{\partial t} + (\mathbf{u} \cdot \nabla)T = \sqrt{\frac{1}{PrRa}}\nabla^2 T. \tag{3}$$

The variables were scaled by the reference length, D_0; velocity, $u_{ref} = (a/D_0)(PrRa)^{1/2}$; pressure, $p_{ref} = pu_{ref}^2$; and time, D_0/u_{ref}. The temperature was scaled as $T = (T^* - T_0)/(T_i - T_0)$, where T^* is the dimensional temperature. A no-slip and no-penetrating condition was applied for the velocity components on all solid walls, and the temperature was $T = 1$ for the plate and $T = 0$ for the enclosure. Although the geometry is left-right symmetric about the vertical centerline ($x/D = 0.0$) for all simulations in this work, we chose to discretize the whole computational domain, instead of half of the domain with the symmetric condition at the vertical centerline, to avoid failure in capturing any possible oscillatory flow behaviors.

The above governing equations were solved by our in-house finite-difference code, which has been employed and well validated in our earlier works for steady-state natural convection in an enclosure [9,18], transient or unsteady mixed convection in an enclosure [19,20], and forced convection across cylinders [21–26]. The numerical details are omitted here for simplicity. In this work, we validated the code through the physical problem of natural convection in an annulus consisting of two concentric circular cylinders at $Ra = 1.71 \times 10^6$, which was experimentally and numerically studied by Kuehn and Goldstein [27] and detailed quantitative benchmark results were provided. This configuration had the same geometric topology as the one in our simulation. The mean equivalent conductivity coefficient was defined as the ratio between the heat transfer rate of the convection case and pure conduction case:

$$k_{eq} = \frac{Nu_{avg,conv}}{Nu_{avg,cond}}, \tag{4}$$

in which the mean Nusselt number was computed as:

$$Nu_{avg} = \frac{1}{\partial\Omega} \int_{\partial\Omega} Nu \, d\partial\Omega, \tag{5}$$

where $\partial\Omega$ is the surface area of the enclosure or plate. The thermal conductivity coefficient reflects the additional heat transfer provided by fluid circulation in addition to conduction and is an indicator of the convection intensity. The value of k_{eq} obtained under various resolutions is given in Table 2. Our results are in good agreement with the benchmark experimental results. The coefficient obtained under various resolutions does not show much difference, i.e., a relative difference of less than 0.2% between the results obtained at the 256 × 128 and 512 × 256 grid, thus the 512 × 256 grid was deemed sufficient for this simulation.

Table 2. Mean equivalent conductivity coefficient obtained under various resolutions at $Ra = 1.71 \times 10^6$ against the numerical results by Kuehn and Goldstein.

Source	Mesh	$k_{eq,i}$	$k_{eq,o}$
Kuehn and Goldstein	16 × 19	3.024	2.973
Present	128 × 64	2.972	2.981
	256 × 128	2.961	2.967
	512 × 256	2.958	2.962

In this work, we performed a grid sensitivity study for the $H \cdot W/D^2 = 0.04$ cases with a square cylinder ($H/D = W/D = 0.2$), a vertically placed plate with $H/D = 0.8$, and a horizontally placed plate with $W/D = 0.8$ using the 256 × 256, 512 × 256, and 512 × 512 grids following the manner described in [28,29]. The computed mean Nusselt numbers on the enclosure surface are given in Table 3. It is seen that the maximum relative difference of the Nusselt number computed under the 256 × 256 and 512 × 256 is relatively large, while the maximum difference between the results of the 512 × 256 and 512 × 512 grid is only about 0.6%. The 512 × 256 grid was found to fine enough and was used in all simulations. The size of the first-layer grid in the wall-normal direction on the plate surface was around $0.0005D$ with clustering at the sharp corners, and was around $0.001D$ on the enclosure surface.

Table 3. Mean Nusselt number $Nu_{avg,o}$ for the $H \cdot W/D^2 = 0.04$ case under different resolutions.

Resolution	$H/D = W/D = 0.2$	$H/D = 0.8$	$W/D = 0.8$
256 × 256	3.965	6.574	5.134
512 × 256	4.102	6.699	5.367
512 × 512	4.125	6.705	5.379

3. Results and Discussion

3.1. Mean Heat Transfer

The heat transfer performance of the plate-enclosure system was first assessed by the mean heat transfer rate. In Figure 2a, the mean Nusselt number is presented for the enclosure. In general, Nu_{avg} monotonically increases with the cross-section area of the plate, and the value for the vertical plate cases is larger than that of the horizontal plate cases. A further inspection on the variations shows that for the vertical plate, Nu_{avg} increases rapidly with H, while the increase with W for the horizontal plate is notably minor, especially for plates with a large cross-section area, e.g., almost constant Nu_{avg} for the $H \cdot W/D^2 = 0.09$ plate, which is nearly the same as the square plate. This mild variation demonstrates that for a horizontal plate with a large cross-section area, the aspect ratio has tiny effects on Nu_{avg}. However, the mean heat transfer rate is still larger for plates with a large W considering that the surface area changes with the aspect ratio. The variation of the mean Nusselt number for the plate is shown in Figure 2b, which was computed using the same approach as in Equation (5). This subfigure demonstrates the efficiency of the heat transfer in terms of the unit surface area of the plate. It is easily demonstrated that with the increases of H or W, both the mean heat transfer rate and surface area increase, while the heat transfer rate per area, i.e., Nu_{avg}, decreases as the increase of the surface area is not so efficient in enhancing heat transfer. It is seen in the subfigure that for a vertical plate, the plate with a smaller cross-section area has a higher Nu_{avg} because of the reduced surface area at the top and bottom where heat transfer is not efficient; however, for a horizontal plate, the increase of W has a negligible effect, which indicates that we can simply enhance heat transfer by employing a thick plate.

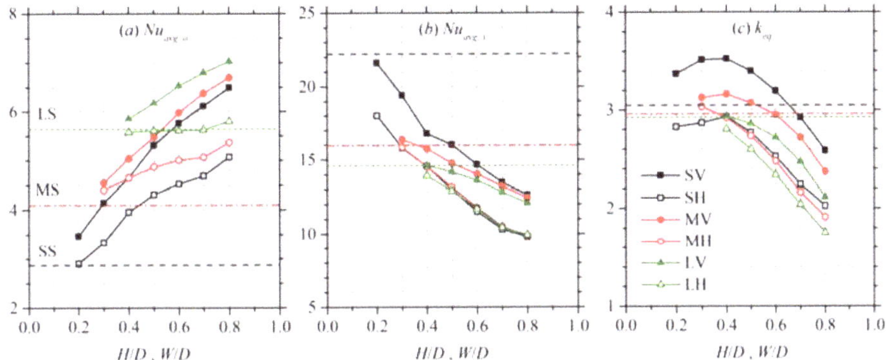

Figure 2. Variations of characteristic quantities: (**a**) mean Nusselt number on the enclosure surface; (**b**) mean Nusselt number on the plate surface; (**c**) mean equivalent conductivity coefficient.

Figure 2c gives the variation of the equivalent thermal conductivity coefficient. Since the fluid is driven by the temperature difference, the convection heat transfer is mainly affected by two factors: The surface area of the plate that heats the adjacent fluid, and the confinement of the plate on the circulation of fluid, which weakens the heat transfer. The combined effect of the two factors determines the convection intensity. For the square geometry ($H = W$), the value of k_{eq} slightly decreases with the cross-section area, which indicates that the large surface area for heating cannot compensate the weakening of the fluid circulation due to the confinement from the plate. It is noticed that the curves in the subfigure exhibit non-monotonic variation with H or W. For the vertical plate, the value of k_{eq} presents an increasing-decreasing pattern with increasing H for plates with $H \cdot W/D^2 = 0.01$ and 0.04, while it monotonically decreases with H for the $H \cdot W/D^2 = 0.09$ geometry. In conclusion, the increase of H or W generally provides less contribution to the convection heat transfer.

3.2. Thermal and Flow Patterns

To further exhibit the effects of the orientation and aspect ratio of the flat plate on the heat transfer mechanism, Figures 3–5 present the distributions of the stream function and temperature in the fluid domain for plates of various cross-section areas. Both distributions are left-right symmetric about the vertical centerline due to the symmetric configuration, thus we gave the isolines of the stream function in the left half of the figure, which coincide with the streamlines, and isotherms in the right half of the enclosure. For the smallest cross-section area, $H \cdot W/D^2 = 0.01$, the plate with a large H has two long surfaces, which could intensely heat the adjacent fluid, thus a primary vortex forms besides the plate, which recirculates within almost the whole enclosure; moreover, the fluid at the center of the vortex is only weakly driven by the temperature difference and the shear stress breaks the vortex into two smaller ones. It is noticed that for a plate with $H/D = 0.8$, the intensity of fluid recirculation to both sides of the plate is strong, and a small and weak vortex that rotates in the clockwise (CW) direction appears at the top of the enclosure. As H decreases, the primary vortex shrinks in size roughly along the vertical direction and the fluid close to the bottom of the enclosure is less affected; the shrinking vortex does not break into multiple ones at its center. The isotherms mainly present a stratification structure. For the horizontal plate, the circulation of fluid within the enclosure is confined by the plate and exhibits complex structures. There is only one primary vortex to the bottom of the plate at W/D = 0.2–0.3; as W increases, the end of the plate separates the primary vortex into two at W/D = 0.4–0.7, with one CW rotating vortex entirely above the plate and one counterclockwise (CCW) rotating vortex beside and below it. For the longest plate at W/D = 0.8, the CCW vortex is separated into two smaller individual ones. Since the fluid below the plate is less driven by the buoyancy, the fluid circulation is strong above the plate and is dominated by the CW vortex. It is seen from the isotherms that for the long horizontal plate (W/D = 0.4–0.8), the thermal plume structure forms at the end whose direction is approximately along the boundary between the CW and CCW vortices.

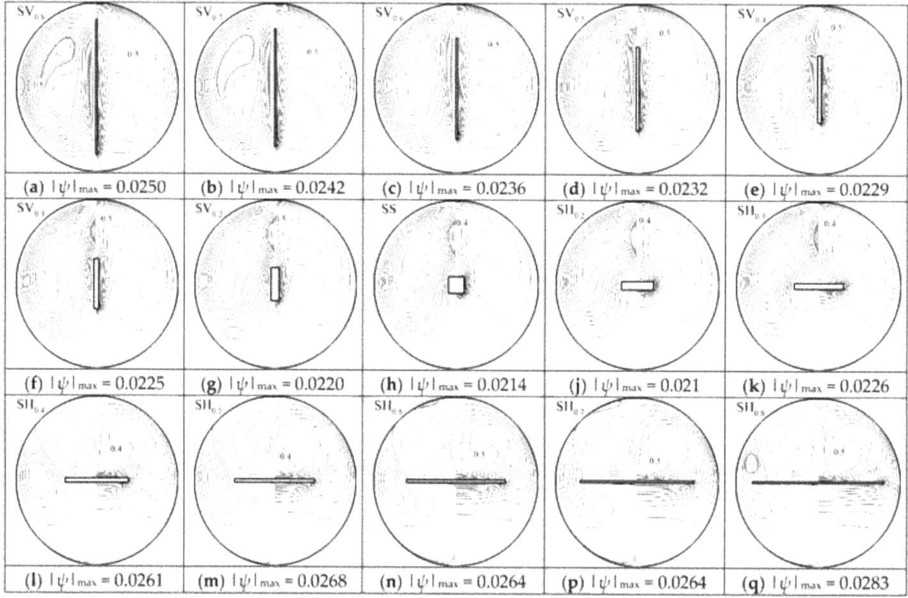

(a) $|\psi|_{max}$ = 0.0250	(b) $|\psi|_{max}$ = 0.0242	(c) $|\psi|_{max}$ = 0.0236	(d) $|\psi|_{max}$ = 0.0232	(e) $|\psi|_{max}$ = 0.0229
(f) $|\psi|_{max}$ = 0.0225	(g) $|\psi|_{max}$ = 0.0220	(h) $|\psi|_{max}$ = 0.0214	(j) $|\psi|_{max}$ = 0.021	(k) $|\psi|_{max}$ = 0.0226
(l) $|\psi|_{max}$ = 0.0261	(m) $|\psi|_{max}$ = 0.0268	(n) $|\psi|_{max}$ = 0.0264	(p) $|\psi|_{max}$ = 0.0264	(q) $|\psi|_{max}$ = 0.0283

Figure 3. Fields of streamlines (left semicircle) and isotherms (right semicircle) for the small-sized plate ($H{\cdot}W/D^2$ = 0.01). The isolines of the stream function are plotted approximately at $\Delta\psi = 0.1|\psi_{max} - \psi_{min}|$; the solid lines are for the counterclockwise rotating vortex and the dashed lines are for clockwise rotating vortex. The isotherms are plotted at ΔT = 0.1.

As the cross-section area increases, the thickness of the plate will also increase for all lengths. The increased thickness has two significant effects on the thermal and flow patterns. The first effect is that the fluid adjacent to the short surface of the plate is more notably heated. It is seen in Figure 4 that the enhanced heating on the fluid generates the CW vortex at the top of the plate at H/D = 0.6–0.8, whose size is larger than that of the $H{\cdot}W/D^2$ = 0.01 cases; as H decreases, the upward circulating flow above the plate is relatively weaker since the plate is far from the top of the enclosure, thus the CW vortex diminishes. Moreover, there is one small CW vortex just above the plate at H/D = 0.3–0.5 because of the separation of upward flow at the sharp corners of the plate. The second effect resulting from the increased thickness is the separation of the primary vortex by the end of the horizontal plate. There is always one CW vortex above the plate, and two CCW vortices in the gap region between the end of the plate and the enclosure, or even three CCW vortices at W/D = 0.8–0.9. The effect of the thick plate on the isotherms is the thermal plume structures at the top sharp corners, which is observable for all horizontal plate cases. The above two effects are more remarkable for the $H{\cdot}W/D^2$ = 0.09 case as shown in Figure 5; the CW vortex above the plate always exists and dominates the circulating flow for the horizontal plate cases, and the thermal plume structure at the sharp corners persists.

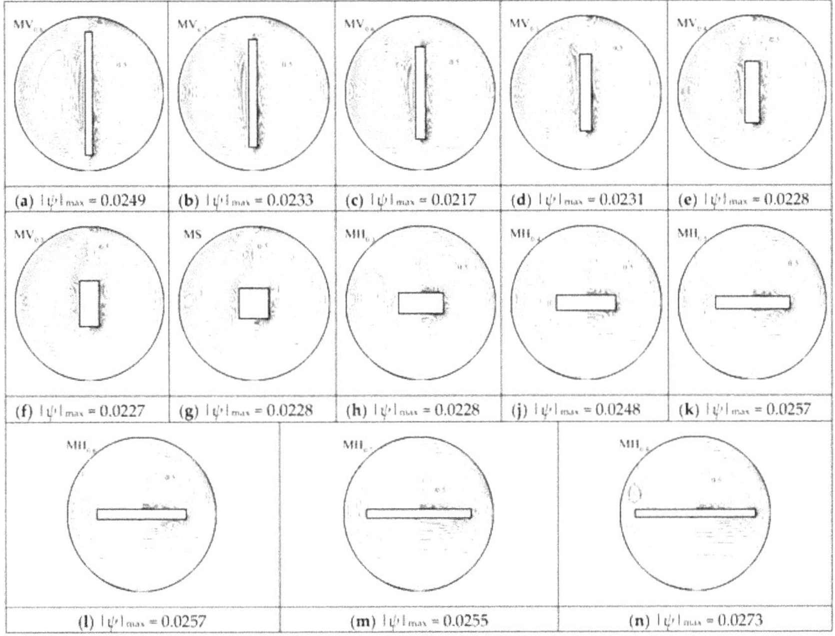

Figure 4. Fields of streamlines (left semicircle) and isotherms (right semicircle) for the middle-sized plate ($H·W/D^2 = 0.04$). The isolines are plotted the same as in Figure 3.

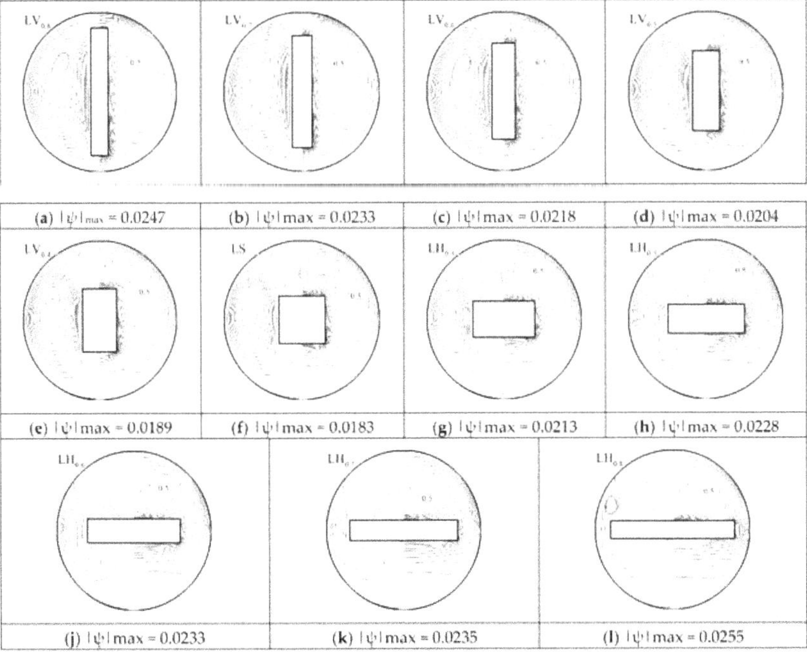

Figure 5. Fields of streamlines (left semicircle) and isotherms (right semicircle) for the middle-sized plate ($H·W/D^2 = 0.04$). The isolines are plotted the same as in Figure 3.

3.3. Local Heat Transfer

The temperature distribution affected by the formation of various vortices would influence the heat transfer on the solid walls and may determine the performance of the plate-enclosure system in certain engineering applications. Figure 6 presents the distribution of the local Nusselt number in the circumferential direction on the wall of the enclosure. Considering the symmetric configuration and temperature field, only the curves for the right half of the enclosure were plotted with the circumferential angle, as given in Figure 1, and the left and right columns of the subfigures denote the vertical and horizontal plate cases, respectively. For the vertical plate, it is obviously seen that the magnitude of the local Nusselt number gradually decreases with the decreasing H for the majority of the enclosure. For the $H \cdot W/D^2 = 0.01$ plate, there is a bump at $\theta = 10°-20°$ for the $H/D = 0.6-0.8$ configurations because of the small gap between the sharp corners of the plate and the enclosure, and it vanishes for $H/D \leq 0.5$ because the end of the plate is relatively far away from the enclosure. It is also noticed that the magnitude of the local Nusselt number at the top of the enclosure does not occur for the longest plate at $H/D = 0.8$ because the fluid above the plate is confined and local heat transfer is conduction dominant, while the circulating flow enhances the heat transfer for plates with a reduced length. The small gap for the longest plate also results in the bump at the bottom of the enclosure. For the plate with a cross-section area of $H \cdot W/D^2 = 0.04$, three differences are observed compared with the results of the $H \cdot W/D^2 = 0.01$ cases. The first is the larger amplitude of the local Nusselt number variation, with H at the top of the enclosure; the second is the bump observed for the $H/D = 0.5$ plate; and the third is the non-monotonic variation of the local Nusselt number, especially in the region of $\theta < 80°$, which is mainly attributed to the clustered isotherms because of the thicker plate. For the plate of $H \cdot W/D^2 = 0.09$, the fluid in the gap to the top of the plate is always dominated by the weak secondary vortex, thus the magnitude of the local Nusselt number is reduced. The thick plate also reduces the gap size and induces the bump on all curves because of the thermal plume structure, and a higher local Nusselt number at the bottom of the enclosure.

For the horizontal plate, the variation of the local Nusselt number with W is more complex because it is dependent on the various vortices. For the $H \cdot W/D^2 = 0.01$ plate at $W/D = 0.2-0.3$, the Nusselt number at the top of the enclosure is large because of the strongly circulating primary vortex. However, as W increases, the magnitude of the Nusselt number abruptly reduces since the secondary vortex is relatively weak and the heat transfer at the enclosure top is not so intense, and a bump is observed on the curve because of the thermal plume; a second bump at $\theta = 95°$ occurs due to the conduction in the small gap. For the $H \cdot W/D^2 = 0.04$ and 0.09 plates, since the secondary vortex and thermal plume exist for all lengths, a local maximum around $\theta = 60°$ is always observed on the curves, which corresponds to the thermal plume. In general, for the horizontal plate, the plate length, W, has a minor effect on the local Nusselt number roughly in the region $\theta < 60°$, while the interaction of CW and CCW vortices and the separation of the primary vortex by the end of the plate in the region $\theta > 60°$ complicate the local flow pattern, thus the effect of W on the local Nusselt number is also more pronounced.

The orientation and aspect ratio of the plate also affect the heat transfer characteristics on the surface of the plate. Figure 7 gives the distribution of the local Nusselt number on the plate, in which the local coordinate, s, is depicted in Figure 1. It is summarized that there is always a peak at the sharp corners of the plate, which can be one order of magnitude higher than that of the flat surface, and the magnitude of the Nusselt number is relatively higher at the bottom of the plate due to the fluid circulation. For the vertical plate, the Nusselt number generally increases with decreasing H, and the differences between the several curves get smaller as the cross-section area increases. For the horizontal plate, the secondary vortex possibly forming above the plate reduces the local Nusselt number in the region, $s < 0.2$, which corresponds to the sharp top corner. As W increases, the magnitude of the Nusselt number for the top and bottom surfaces nearly monotonically decreases because of the confinement of the plate on the fluid circulation.

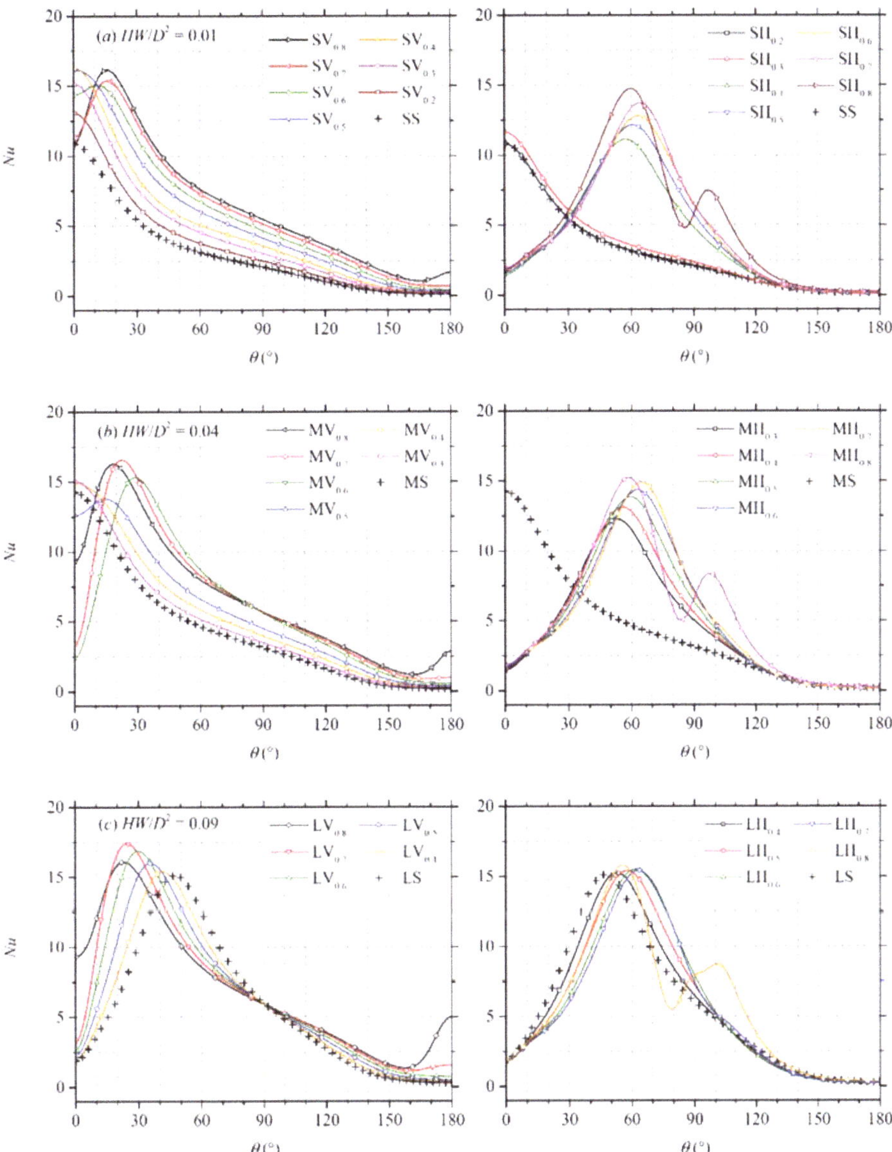

Figure 6. Circumferential distribution of the local Nusselt number on the enclosure surface.

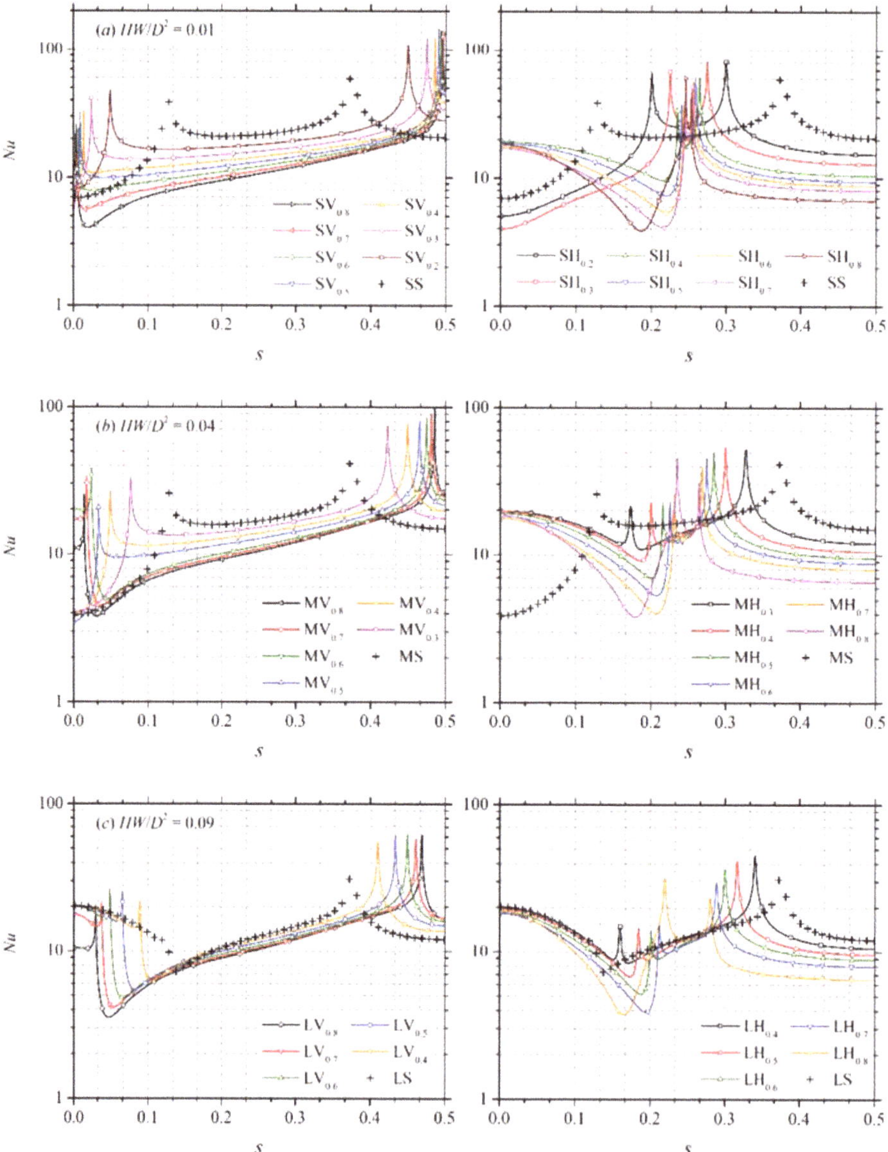

Figure 7. Circumferential distribution of the local Nusselt number on the plate surface.

3.4. Synergy Principle Analysis

The enhancement of heat transfer contributed by convection with respect to pure conduction is a result of the circulation of fluid within the enclosure. Quantitative analysis of the enhancement is significant in understanding the related physics. Since the fluid moves in parallel with the solid wall in the boundary layer region, the fluid-solid heat transfer is realized almost by pure conduction. However, in the fluid domain outside of the boundary region, the intensity of heat transfer is dependent on the direction of flow considering the non-uniform temperature distribution, i.e., the alignment of the vectors of the velocity and temperature gradient. The heat transfer is enhanced if the velocity vector is aligned in the direction with the temperature gradient in which the fluid circulation could effectively mix the hot and cold fluid and increase the local heat transfer rate. Guo et al. [30] proposed the field synergy principle:

$$\mathbf{U} \cdot \nabla T = |\mathbf{U}||\nabla T| \cos \beta, \tag{6}$$

in which β represents the synergy angle between the two vectors. The local heat transfer can be enhanced if the two vectors are synergized.

We computed the synergy coefficient, $\cos\beta$, for this problem and present its distribution in Figure 8. In general, for the vertically placed plate, the value of $\cos\beta$ is almost zero at the solid walls, which indicates that the local heat transfer is conduction dominant. The enhancement of heat transfer from convection mainly occurs in the central region of the fluid domain and exhibits a zigzag structure with a relatively large magnitude, thus the local heat transfer from the mixing of cold and hot fluid is intense. The magnitude of the synergy coefficient to the top of the plate is also large, especially for plates with a larger cross-section area, which is attributed to the formation of the CW vortex and thermal plume structure. It is also noticed that due to the existence of the CW vortex, the direction of the fluid motion is reversed from upward for the $H \cdot W/D^2 = 0.01$ case to downward for the $H \cdot W/D^2 = 0.04$ and 0.09 cases, thus the magnitude for the local synergy coefficient is also changed. A similar distribution pattern is also observed for the horizontally placed plate. However, due to the formation of multiple CW and CCW vortices in the gap region between the end of the plate and the enclosure, the distribution of $\cos\beta$ is even more complex. Since the CW vortex dominates the flow above the plate, the magnitude of $\cos\beta$ is large at the vortex center and its boundary where the thermal plume appears. The effect of the plate length is noticeable mainly in the gap region, where the CCW vortex is separated by the plate; it has only one local maximum at the center of the vortex for the relatively short plate where there is one CCW vortex, while two local maxima are observed as the CCW vortex is separated by the long plate into two.

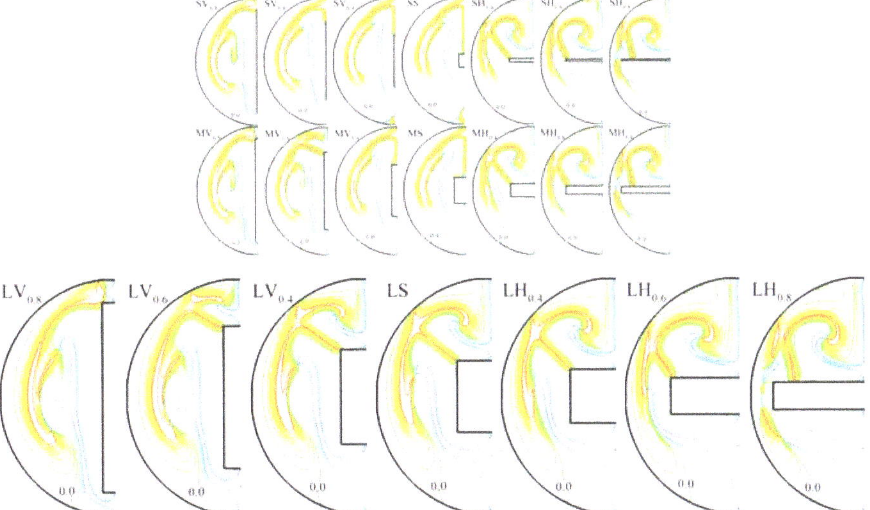

Figure 8. Field of synergy coefficient plotted at $\cos\beta = [-1.0, 1.0]$ with $\Delta\cos\beta = 0.1$. The solid and dashed lines respectively denote the positive and negative values.

4. Conclusions

This work performed a numerical investigation on the natural convection in a circular enclosure with an internal flat plate of a constant cross-section area but different orientation and aspect ratio.

For this physical problem, we first explored the trade-off between the height and width of the plate under the constraint of a fixed cross-section area by considering the effects of the heating surface area and fluid circulating confinement, since the heat transfer effectiveness is greatly dependent on these two factors. The numerical results led to the following conclusions:

(1) The vertically placed plate produces significant convection heat transfer with a higher mean Nusselt number than the horizontal plate. However, the overall heat transfer rate does not vary much with the width of the plate for the horizontally placed plate cases.

(2) For the vertically placed plate, there is always one primary vortex to each side of the plate, and one secondary vortex may form above the plate as the width increases, especially for plates with larger cross-section areas. For the horizontally placed plate with a large width, the primary vortex can be separated by the end of the plate into two, and a secondary vortex forms above the plate.

(3) The magnitude of the local heat transfer rate on the enclosure surface is higher as the circulating flow impinges on the surface, and it is normally maximum at the circumferential position of the thermal plume. The secondary vortex above the plate actually reduces the local heat transfer rate at the top of the enclosure.

(4) A field synergy analysis revealed that the contribution of fluid circulation to the convection heat transfer mainly appears at the center of the vortices, at the boundary of neighboring vortices where the thermal plume forms, and above the plate.

The present work performed more of a theoretical study on the natural convection in an enclosure with a specific geometric configuration. However, considering that the flat plate geometry is widely used in engineering applications, the findings and conclusions could provide a useful guidance or reference for other applications in choosing the proper geometry and orientation to maximize heat transfer effectiveness.

Author Contributions: Conceptualization, A.W. and W.Z.; Data curation, A.W.; Formal analysis, A.W.; Funding acquisition, L.Z. and W.Z.; Investigation, A.W., C.Y., Y.W., L.Y. and Y.Y.; Methodology, W.Z.; Project administration, L.Z. and W.Z.; Resources, L.Z. and W.Z.; Software, W.Z.; Supervision, L.Z. and W.Z.; Validation, A.W.; Visualization, A.W.; Writing—Original draft, A.W.; Writing—Review and editing, W.Z.

Funding: This research was funded by National Natural Science Foundation of China, grant number 51706205 51606170 and 51676173.

Conflicts of Interest: The authors declare no conflict of interest.

References

1. Alteç, Z.; Kurtul, Ö. Natural convection in tilted rectangular enclosures with a vertically situated hot plate inside. *Appl. Therm. Eng.* **2007**, *27*, 1832–1840. [CrossRef]

2. Wang, X.; Shi, D.; Li, D. Natural convective flow in an inclined lid-driven enclosure with a heated thin plate in the middle. *Int. J. Heat Mass Trans.* **2012**, *55*, 8073–8087. [CrossRef]

3. Öztop, H.F.; Dagtekin, I.; Bahloul, A. Comparison of position of a heated thin plate located in a cavity for natural convection. *Int. Commun. Heat Mass Trans.* **2004**, *31*, 121–132. [CrossRef]

4. Tasnim, S.H.; Collins, M.R. Suppressing natural convection in a differentially heated square cavity with an arc shaped baffle. *Int. Commun. Heat Mass Trans.* **2005**, *32*, 94–106. [CrossRef]

5. Wang, Q.W.; Yang, M.; Tao, W.Q. Natural convection in a square enclosure with an internal isolated vertical plate. *Warme Stoffubertrag* **1994**, *29*, 161–169. [CrossRef]

6. Abdul Hakeem, A.K.; Saravanan, S.; Kandaswamy, P. Buoyancy convection in a square cavity with mutually orthogonal heat generating baffles. *Int. J. Heat Fluid Flow* **2008**, *29*, 1164–1173. [CrossRef]

7. Kandaswamy, P.; Abdul Hakeem, A.K.; Saravanan, S. Internal natural convection driven by an orthogonal pair of differentially heated plates. *Comput. Fluids* **2015**, *111*, 179–186. [CrossRef]

8. Singh, P.; Liburdy, J.A. Effect of plate inclination on natural convection from a plate to its cylindrical enclosure. *ASME J. Heat Transf.* **1986**, *108*, 770–775. [CrossRef]

9. Zhang, W.; Wei, Y.; Chen, X.; Dou, H.-S.; Zhu, Z. Partitioning effect on natural convection in a circular enclosure with an asymmetrically placed inclined plate. *Int. Commun. Heat Mass Trans.* **2018**, *90*, 11–22. [CrossRef]

10. Dagtekin, I.; Öztop, H.F. Natural convection heat transfer by heated partitions within enclosure. *Int. Commun. Heat Mass Trans.* **2001**, *28*, 823–834. [CrossRef]

11. Nag, A.; Sarkar, A.; Sastri, V.M.K. Natural convection in a differentially heated square cavity with a horizontal partition plate on the hot wall. *Comput. Method Appl. Mech. Eng.* **1993**, *110*, 143–156. [CrossRef]

12. Tasnim, S.H.; Collins, M.R. Numerical analysis of heat transfer in a square cavity with a baffle on the hot wall. *Int. Commun. Heat Mass Trans.* **2004**, *31*, 639–650. [CrossRef]

13. Sun, C.; Yu, B.; Öztop, H.F.; Wang, Y.; Wei, J. Control of mixed convection in lid-driven enclosures using conductive triangular fins. *Int. J. Heat Mass Trans.* **2011**, *54*, 894–909. [CrossRef]

14. Wei, Y.; Yang, H.; Dou, H.-S.; Lin, Z.; Wang, Z.; Qian, Y. A novel two-dimensional coupled lattice Boltzmann model for thermal incompressible flows. *Appl. Math. Comput.* **2018**, *339*, 556–567. [CrossRef]

15. Lun, Y.X.; Lin, L.M.; Zhu, Z.C.; Wei, Y.K. Effects of vortex structure on performance characteristics of a multiblade fan with inclined tongue, P.I. *Mech. Eng. Part A* **2019**, *233*, 1007–1021. [CrossRef]

16. Yang, H.; Yu, P.; Xu, J.; Ying, C.; Cao, W.; Wang, Y.; Zhu, Z.; Wei, Y. Experimental investigations on the performance and noise characteristics of a forward-curved fan with the stepped tongue. *Meas. Control* **2019**, in press. [CrossRef]

17. Tao, J.; Lin, Z.; Ma, C.; Ye, J.; Zhu, Z.; Li, Y.; Mao, W. An experimental and numerical study of regulating performance and flow loss in a V-port ball valve. *J. Fluids Eng.* **2020**, *142*, 021207. [CrossRef]

18. Wang, Y.; Chen, J.; Zhang, W. Natural convection in a circular enclosure with an internal cylinder of regular polygon geometry. *AIP Adv.* **2019**, *9*, 065023. [CrossRef]

19. Zhang, W.; Wei, Y.; Dou, H.-S.; Zhu, Z. Transient behaviors of mixed convection in a square enclosure with an inner impulsively rotating circular cylinder. *Int. Commun. Heat Mass Trans.* **2018**, *98*, 143–154. [CrossRef]

20. Yang, H.; Zhang, W.; Zhu, Z. Unsteady mixed convection in a square enclosure with an inner cylinder rotating in a bi-directional and time-periodic mode. *Int. J. Heat Mass Trans.* **2019**, *136*, 563–580. [CrossRef]

21. Zhang, W.; Chen, X.; Yang, H.; Liang, H.; Wei, Y. Forced convection for flow across two tandem cylinders with rounded corners in a channel. *Int. J. Heat Mass Trans.* **2019**, *130*, 1053–1069. [CrossRef]

22. Zhang, W.; Yang, H.; Dou, H.-S.; Zhu, Z. Forced convection of flow past two tandem rectangular cylinders in a channel. *Numer. Heat Tr. Part A* **2017**, *72*, 89–106. [CrossRef]

23. Zhang, W.; Li, X.; Zhu, Z. Quantification of wake unsteadiness for low-Re flow across two staggered cylinders, P.I. *Mech. Eng. Part C* **2019**, *233*, 6892–6909. [CrossRef]

24. Zhang, W.; Dou, H.-S.; Zhu, Z.; Li, Y. Unsteady characteristics of low-Re flow past two tandem cylinders. *Theor. Comput. Fluid Dyn.* **2018**, *32*, 475–493. [CrossRef]

25. Zhang, W.; Samtaney, R. Effect of corner radius in stabilizing the low-Re flow past a cylinder. *J. Fluids Eng.* **2017**, *139*, 121202. [CrossRef]

26. Zhang, W.; Yang, H.; Dou, H.-S.; Zhu, Z. Flow unsteadiness and stability characteristics of low-Re flow past an inclined triangular cylinder. *J. Fluids Eng.* **2017**, *139*, 121203. [CrossRef]

27. Kuehn, T.H.; Goldstein, R.J. An experimental and theoretical study of natural convection in the annulus between horizontal concentric cylinders. *J. Fluid Mech.* **1976**, *74*, 695–719. [CrossRef]

28. Siddique, W.; El-Gabry, L.; Shevchuk, I.V.; Hushmandi, N.B.; Fransson, T.H. Flow structure, heat transfer and pressure drop in varying aspect ratio two-pass rectangular smooth channels. *Heat Mass Transf.* **2012**, *48*, 735–748. [CrossRef]

29. Siddique, W.; Shevchuk, I.V.; El-Gabry, L.; Hushmandi, N.B.; Fransson, T.H. On flow structure, heat transfer and pressure drop in varying aspect ratio two-pass rectangular channel with ribs at 45°. *Heat Mass Transf.* **2013**, *49*, 679–694. [CrossRef]

30. Guo, Z.Y.; Tao, W.Q.; Shah, R.K. The field synergy (coordination) principle and its applications in enhancing single phase convective heat transfer. *Int. J. Heat Mass Transf.* **2005**, *48*, 1797–1807. [CrossRef]

Article

Thermo-Diffusion and Multi-Slip Effect on an Axisymmetric Casson Flow over a Unsteady Radially Stretching Sheet in the Presence of Chemical Reaction

Faraz Faraz [1,*], Syed Muhammad Imran [2], Bagh Ali [1] and Sajjad Haider [3]

[1] Department of Applied Mathematics, School of Science, Northwestern Polytechnical University, Dongxiang Road, Chang'an District, Xi'an 710129, China; baghalisewag@mail.nwpu.edu.cn

[2] Department of Mathematics Govt postgraduate Gordon College, Rawalpindi 46000, Pakistan; syedsiim@gmail.com

[3] College of Applied Science, Beijing University of Technology, Beijing 100124, China; Sajjadsaleem266@hotmail.com

* Correspondence: faraz313pak@mail.nwpu.edu.cn or faraz313pak@hotmail.com

Received: 17 October 2019; Accepted: 8 November 2019; Published: 14 November 2019

Abstract: The objective of this article is to investigate the impacts of thermo-diffusion effect on unsteady axisymmetric Casson flow over a time-dependent radially stretching sheet with a multi-slip parameter and the force of chemical reaction. We employed an established similarity transformation to this non-linear partial differential system to convert it into a system of ordinary differential equations. The numerical results are attained for this system by using KELLER-BOX implicit finite difference scheme. It has great reliability and accuracy even a very short time period for computational simulation. The impacts of influential flow parameters on fluid flow are sketched through graphs and the numerical results are thoroughly argued. The temperature, velocity and wall concentration control parameters are analyzed. (i) It is witnessed that chemical reaction is not favorable to enhance the velocity profile. (ii) Multi-slip parameters vary inversely with velocity profile. (iii) The fluid concentration in its boundary layer decreases with the increase of heavier species, the parameter of the reaction rate and the exponent of power law for fluids having Prandtl number = 10.0, 15.0, 20.0 and 25.0. Moreover, the skin-friction-coefficient factor and Nusselt-number are compared with the published work. A strong numerical solution agreement is being observed.

Keywords: multi-slip; Keller-Box technique; casson fluid; thermo-diffusion; axisymmetric flow

1. Introduction

The knowledge of non-Newtonian fluids has great importance for their characteristics and remarkable applications in industrial, medical products, and procedures to the researcher. These all non-Newtonian type fluids have non-linear relation between stress and strain, whereas Newtonian fluid model has a linear relations mode. Investigations of flow field and individualities in these fluids are completely different as compare to Newtonian fluids. The Casson fluid model is popular for good explanation of non-Newtonian fluids and their behavior, especially flow curves for blood. It is recordable convincing fluid model because of important useful implications in our daily life as in bio medical field and polymer processing. For practical purposes, it provides a convenient means for evaluating the two characteristics; Cason viscosity and the apparent yield stress. A great number of investigations have described about this chemical fields. In species research of mass and heat transformations with chemical reactions are of extensive importance in specially hydrometallurgical and chemical at the industrial level. Last past few decades, a basic "Penetration Theory" "Highie 1935" had been extensively practically applied to the time dependent diffusional problems without and with chemical reactions. As long we ascertain all about the results with chemical reactions were found for the case of mid-infinite bodies of fluids, even though physically absorptions into a determinate based film were measured. Several interesting phenomena were considered for analysis of mass transformations liminal with forced convection and chemical reactions [1]. For the work about vapor-deposition chemically fundamental results are obtained [2] . In diffusion model the chemical effect on browning motion was carried out by [3]. Some of the others investigation was studied by [4–7].

Until now, in the absence of Soret and Dufour impacts, all the above studies have been conducted. In a flowing fluid, heat and mass transfer occurring simultaneously results in a complicated relationship between the fluxes and the fluid's flowing existence. Energy diffusion can be produced not only by gradients of temperature, but also by gradients of composition. The temperature gradients that result in Soret (thermo-diffusion) effect can create mass fluxes. At the other hand, the effect of the energy fluxes causedby the gradients of composition is called Dufour (diffusion-thermo). Such fluxes play an important role when there is a density difference in the flow regime.

The fluids of the boundary layer flow due to stretching /shrinking surfaces is a significant kind of flow occurring in engineering and chemical industries flow processes. These include paper production, liquid metal, glass fiber, and polymer sheet synthesis. The manufacture of non-newtonian fluids, including lubricants, physiological liquids, paints, colloidal liquids, biological liquids, biopolymer, and foodstuff, plays an important role in our daily lives. Bagh et al. [8] examined the influence of multiple slip-on non-newtonian fluids and they described that the velocity profile decline due to increasing in the hydrodynamic slip. Raza [9] analyzed the Casson fluid flow over a sheet and examined the radiation effects on temperature. Ashraf and coauthors [10] have investigated the micro-polar fluid flow toward a shrinking surface and also studied the radiation effects on thermal conductivity. Daniel et al. [11] studied the numerical solution of mixed convection magnetohydrodynamic flow over a sheet. Dhanai et al. [12] Several Magneto-hydro-dynamics (MHD) heat transfer fluid solutions were achieved with viscous dissipation. The study of the unsteady axisymmetric flow of non-Newtonian fluid over a radially stretch sheet has considered by Shahzad et al. [13]. They also studied the radiation effects on the thermal boundary layer. Ashraf et al. [14] examined the magnetohydrodynamic flow and heat transfer in a micro-polar fluid using a stretchable disk. Azeem et al. [15] analyzed the heat transfer of an axisymmetric viscous fluid over a nonlinear radially stretching sheet.

The study of free convection flow is important in the electronics cooling process and heat exchangers etc. Chen [16] examined the laminar mixed convection flow over a continuously stretching sheet. Numerous researchers have been occupied with investigating the mixed convection flow of non-Newtonian fluids [17–19]. Bhargava et al. [20] have analyzed the free convection flow of magnetohydrodynamic micro-polar fluid. Elahi et al. [21] described the numerical solution of mixed convection heat transfer over a stretching sheet. Asmat et al. [22] studied the effect of thermal radiation on velocity and temperature over a stretching porous sheet. Hayat et al. [23] considered on magnetohydrodynamic the flow of non-Newtonian nano-fluid flow with the convective condition. They investigated the slip effects in the MHD flow of non-Newtonian by a stretching surface. They also found radiation effects on velocity, temperature and concentration profile.

Baag et al. [24] have studied the stagnation point on magnetohydrodynamic non-newtonian fluid subject to the chemical reaction and heat source. Singh [25] examined the effects of viscous on free convection non-newtonian fluid in the presence of chemical reaction. Mabood et al. [26] discussed steady non-Newtonian fluid with a chemical reaction through a porous medium. Hayat et al. [27] are discussed non-Newtonian fluid with chemical aspects and they investigated a numerical solution. Seth et al. [28] examined chemically reacting nanofluid over a permeable vertical plate.

Motivated by the above-mentioned studies in literature and a wide range of their applications, the thermo-diffusion and multi-slip effects on an axisymmetric Casson flow over a time-dependent radially stretching sheet in the presence of chemical reaction is presented which has not been discussed yet. The focal point of the current study is to extend the recently published work of Azeem et al. [15].

The governing nonlinear PDEs are transformed into a set of highly nonlinear ODEs with the aid of suitable similarity transformations and the nonlinear coupled ODEs are solved numerically with most popular Keller-Box technique. The effects of magnetic parameter M, Dufour parameter D_s, Schmidt number parameter Sc, chemical reaction parameter R_0, Soret parameter D_f, Prandtl number Pr, slip parameters (δ_1, δ_2), suction/injection parameter S, Unsteadiness parameter α, thermal buoyancy λ, and Casson parameter β on the fluid velocity, temperature, and concentration functions are examined in detail. Additionally, a comparison is made for the skin friction coefficient and Nusselt number. Good agreement is established which further authenticates the validity of our results.

2. Mathematical Formulation

Let us consider a steady magnetohydrodynamic flow of incompressible viscous flow with thermo-diffusion are included over a radially stretching sheet, the sheet is placed at $z = 0$, and is examined in the presence of chemical reaction effects. The flow of conducting fluid is assumed to be linear along the radial direction $U_w(r) = \frac{ar}{1-ct}$, where a is a dimensional constant. Where T_w is the wall temperature, T_∞ is the ambient temperature respectively. (see Figure 1). It is supposed that the $B(r) = B_0 r$ variable magnetic field intensity acts along z-direction normal to the sheet. Under the above conditions, the governing equations of continuity, momentum conservation, and conservation can be expressed as (see [13,29]):

$$\frac{\partial u}{\partial r} + \frac{u}{r} + \frac{\partial w}{\partial z} = 0, \tag{1}$$

$$\frac{\partial u}{\partial t} + u\frac{\partial u}{\partial r} + w\frac{\partial u}{\partial z} = v_f\left(1 + \frac{1}{\beta}\right)\frac{\partial^2 u}{\partial z^2} - \frac{\sigma B^2(r)u}{\rho_f} + g\beta_T(T - T_\infty) \tag{2}$$

$$\frac{\partial T}{\partial t} + u\frac{\partial T}{\partial r} + w\frac{\partial T}{\partial z} = \alpha\frac{\partial^2 T}{\partial z^2} + D_{TC}\frac{\partial^2 C}{\partial^2 z} \tag{3}$$

$$\frac{\partial C}{\partial t} + u\frac{\partial C}{\partial r} + w\frac{\partial C}{\partial z} = D\frac{\partial^2 C}{\partial z^2} - R^*(r,t)(C - C\infty) + D_{CT}\frac{\partial^2 C}{\partial^2 z} \tag{4}$$

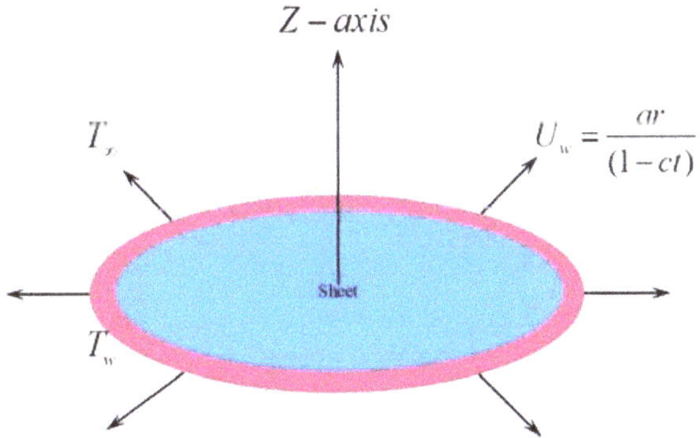

Figure 1. thermo-diffusion flow diagram.

The velocity vector of flow is $v = v(u,v)$, where u and v are component of velocity towards r and z directions respectively. σ, v_f, ρ_f, α, D, D_{CT}, D_{TC}, and R^* are electrical conductivity, kinetic viscosity, the viscosity of fluid, solutal, Soret, Dufour diffusities, and chemical reaction respectively, The corresponding boundary conditions are (see [13,29]):

$$u = U_w + U_s, w = W_0, T = T_w = T_\infty + \frac{br}{1-ct} + T_s, C = C_w \ at \quad z = 0 \tag{5}$$

$$u \to 0, T \to \infty, C \to \infty \ as \quad z \to \infty, \tag{6}$$

where $U_s = D_1 \frac{du}{dr}$ is the velocity slip, D_1 is the velocity slip factor, $T_s = D_2 \frac{dT}{dr}$ is the thermal slip, D_2 is the temperature slip factor, and $W_0 = -2(\frac{\nu U_w}{r})^{\frac{1}{2}}$ denotes the suction/injection of mass transfer rate at the surface. Here a > 0, b > 0, and c > 0 are constant having dimension 1/time (t), where t stand for the time such that product ct < 1 (see [29]).

The Equations (1)–(4), We consider the similarity transformations stated as (see [29]):

$$\eta = \frac{z}{r} Re_r^{\frac{-1}{2}}, w = -2U_w Re_r^{\frac{-1}{2}} f(\eta), u = U_w f'(\eta), \theta(\eta) = \frac{T - T_\infty}{T_w - T_\infty}, \text{ and } \phi(\eta) = \frac{C - C_\infty}{C_w - C_\infty} \tag{7}$$

In view of Equation (7), the system of partial differential Equations (2)–(5) transform into the following system of coupled and non-linear ODE's:

$$(1 + \frac{1}{\beta}) \frac{d^3 f}{d\eta^3} - M \frac{df}{d\eta} + 2f \frac{d^2 f}{d\eta^2} - (\frac{df}{d\eta})^2 - \alpha[\frac{df}{d\eta} + \frac{1}{2}\eta \frac{d2f}{d\eta^2}] + \lambda\theta = 0, \tag{8}$$

$$(\frac{1}{Pr}) \frac{d^2\theta}{d\eta^2} - \theta \frac{df}{d\eta} + 2f \frac{d\theta}{d\eta} + D_s \frac{d^2\phi}{d\eta^2} - \alpha[\theta + \frac{1}{2}\eta \frac{d\theta}{d\eta}] = 0, \tag{9}$$

$$(\frac{1}{Sc}) \frac{d^2\phi}{d\eta^2} - \phi \frac{df}{d\eta} + 2f \frac{d\phi}{d\eta} - \alpha[\phi + \frac{1}{2}\eta \frac{d\phi}{d\eta}] + D_f \frac{d^2\theta}{d\eta^2} - R_0\phi = 0, \tag{10}$$

and the transformed boundary conditions Equations (5) and (6) are:

$$f(\eta) = S, \frac{df(\eta)}{d\eta} = 1 + \delta_1 \frac{d^2 f(\eta)}{d\eta}, \theta(\eta) = 1 + \delta_2 \frac{d\theta(\eta)}{d\eta}, S(\eta) = 1, \phi(\eta) = 1, \quad at \ \eta = 0, \tag{11}$$

$$\frac{df}{d\eta}(\infty) \to 0, \quad \theta(\infty) \to 0, \quad \phi(\infty) \to 0, \ at \quad \eta \to \infty \tag{12}$$

where primes represent differentiation w.r.t the variable η. The parameters in Equations (8)–(10) are described as:

$$M = \frac{\sigma B^2}{\rho_f a}, Pr = \frac{\nu_f \rho c_p}{\kappa}, D_s = \frac{D_{TC}(C_w - C_\infty)}{\nu_f(T_w - T_\infty)}, D_f = \frac{D_C T(T_w - T_\infty)}{\nu(C_w - C_\infty)}, \alpha = \frac{a}{c}, R_0 = \frac{(1 - ct)^2}{a},$$

where M is magnetic parameter, Pr is define as the Prandtl number, D_s is the Dufour parameter, D_f is the Soret parameter, Sc is determine as the Schmidt number, and R_0 is the chemical reaction term.

The interested physical quantities are coefficient of skin-friction C_f, local nusselt number Nu, and Sherwood number Sh defined as:

$$C_f = \frac{\tau_w}{\frac{1}{2}\rho U^2}, \quad Nu = \frac{r q_w}{K(T_w - T_\infty)}, \quad \text{and} \quad Sh = \frac{L q_m}{D(C_w - C_\infty)} \tag{13}$$

whereas the skin-friction coefficient τ_w, the heat and mass transformation from the sheet q_w and q_m are follow:

$$\tau_w = \mu \frac{\partial u}{\partial z}|_{z=0}, \quad q_w = -\kappa \frac{\partial T}{\partial z}|_{z=0}, \quad \text{and} \quad q_m = -D \frac{\partial C}{\partial z}|_{z=0} \tag{14}$$

The dimension free variables explained in Equation (7) and these quantities becomes as:

$$Re_r^{\frac{1}{2}} C_f = (1 + \frac{1}{\beta})f''(0), \quad Re_r^{\frac{-1}{2}} Nu = -\theta'(0), \quad \text{and} \quad Re_r^{\frac{-1}{2}} Sh = -\phi'(0) \tag{15}$$

where $Re_r = \frac{r U_w}{\nu}$ is the local Reynolds number based on the radially stretching velocity $U_w = \frac{ar}{1 - ct}$.

3. Results and Discussion

The main goal of the proposed study was to define the role of mass transformation factor, heat transformation factor, chemical reaction and thermal radiation factors in the time-dependent axisymmetric boundary layer MHD flow of Casson fluid if multiple-slip,and thermo-diffusion effects are employed over a stretching surface.

The control model Equations (8)–(10) with boundary conditions Equations (11) and (12) were solved numerically by Keller-Box finite difference method. The values of the velocity, temperature, and concentration profiles are analyzed in the current section using the numerical technique. We have graphically discussed the influence of these profiles on various parameters such as Casson, magnetic, Prandtl number, Dufour parameter, Soret number, chemical reaction, Schidmt number, unsteadiness, buoyancy, hydrodynamic, suction/injection parameter, and thermal slips. In order to validate the numerical method in Table 1, presents the comparison of our work that of Azeem et al. [29] and an excellent correlation is achieved which shows the authenticity of numerical solutions.

Table 1. Present results are Compared with [29] of $-f''(0)$ and $-\theta'(0)$ for various values of α, S and Pr.

α	S	Pr	Azeem et al. [13]		KBM (Present Results)	
			$-f''(0)$	$-\theta'(0)$	$-f''(0)$	$-\theta'(0)$
0.5	−1.0	1.0	0.620400	0.620400	0.620436	0.620436
	−0.5		0.887200	0.887200	0.887247	0.887247
	0.0		1.308999	1.308999	1.308670	1.308670
	0.5		1.907999	1.907999	1.907973	1.907973
	1.0		2.655999	2.655999	2.655591	2.655591
0.0	0.5	1.0	1.798999	1.798999	1.798668	1.798668
0.5			1.907999	1.907999	1.907973	1.907973
1.0			2.016999	2.016999	2.016665	2.016665
0.5	0.5	0.5	1.907999	1.119999	1.907973	1.118889
		0.7	1.907999	1.450000	1.907973	1.467003
		1.0	1.907999	1.907999	1.907973	1.907973

Figure 2a,b depict the influence of Casson parameter at velocity, temprature and concentration profile. The velocity profile decreases with an increase in Casson fluid parameter (β) but opposite behavior is observed in Figure 2a,b for the temperature and concentration functions. Figure 3a,b shows the influence of the magnetic parameter M on the free dimension velocity, temperature, and concentration profiles, It is noticed that the velocity distribution decline due to increment in the magnetic field. It is clearly seen in Figure 3a increase in the value of M slow down the movementum and hence a decline in radial velocity. Temperature, and concentration profiles are observed increases near to the boundary wall. The escalation in the values of magnetic causes increment in thermal and solutal boundary thickness. Physically, the magnetic parameter produced Lorentz force which slows down the motion of the fluid.

In Figure 4a,b the influence of dimensionless unsteady parameter (α) on the velocity, temperature and concentration profiles is shown. It is found that the velocity, temperature, and concentration profiles decline as an increasing values of the unsteady parameter (α) whereas the temperature profile is an increasing function near the boundary.

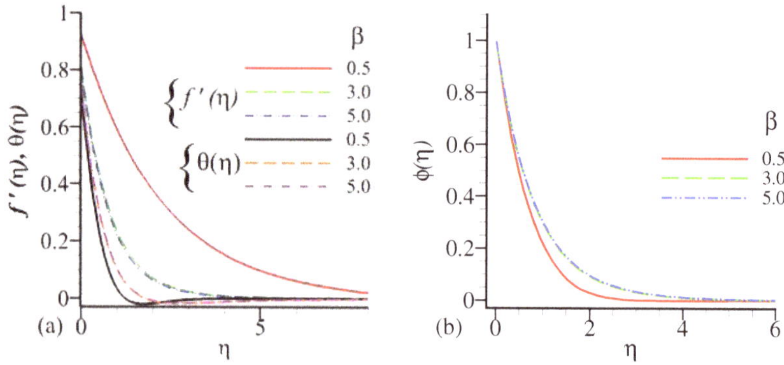

Figure 2. Influence of β on velocity profile f', temperature profile θ, and concentration profile ϕ against η.

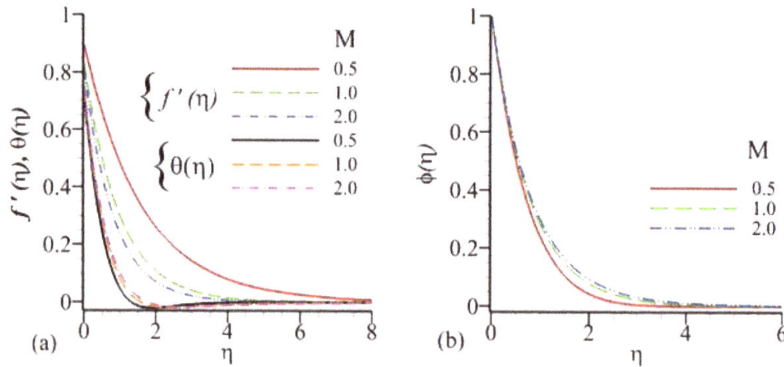

Figure 3. Influence of M on velocity profile f', temperature profile θ, and concentration profile ϕ against η.

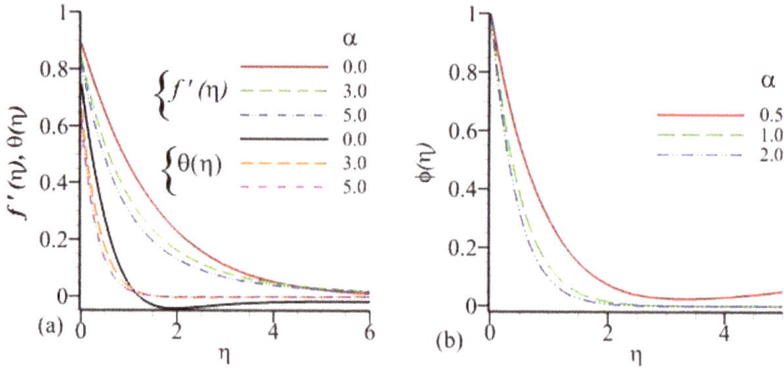

Figure 4. Influence of α on velocity profile f', temperature profile θ, and concentration profile ϕ against η.

Figure 5a,b illustrates the profiles of dimension-less velocity, temperature, and concentration for different values of buoyancy (λ). It is observed that velocity increases whereas the concentration profile and temperature decreased with increasing values of buoyancy (λ) (see Figure 5a,b. The extra force is added in fluid due to buoyant that's why the velocity profile enhance.

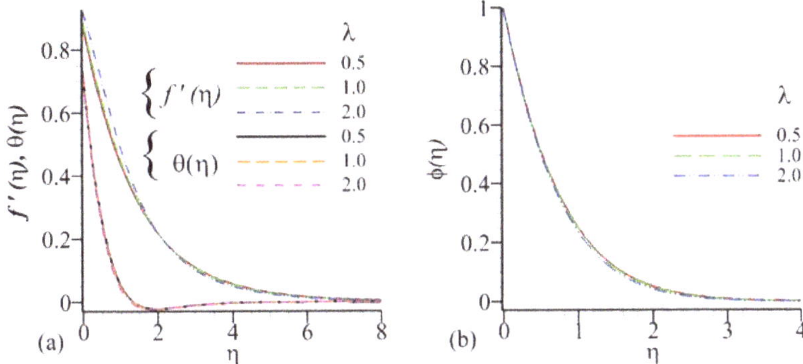

Figure 5. Influence of λ on velocity profile f', temperature profile θ, and concentration profile ϕ against η.

Figure 6a,b depicts radial velocity and temperature profiles in which the boundary layer thickness is reduced near the wall with the rise in Prandtl number Pr. Prandtl number effect on concentration profile slightly differs form radially stretching velocity, and temperature profiles as we can see in Figure 6b, which shows that it is slow rising effect away to the boundary and decreases closed to the boundary layer with rising in Prandtl number. Casson fluids include in the present investigation it has determined with great viscosity therefore, the Prandtl number is used to increase the rate of cooling in conducting flows.This is due to the fact that Pr number is defined as the ratio between momentum and thermal diffusivity. In present investigation $Pr = 25$ number is very suitable for cooling purposed.

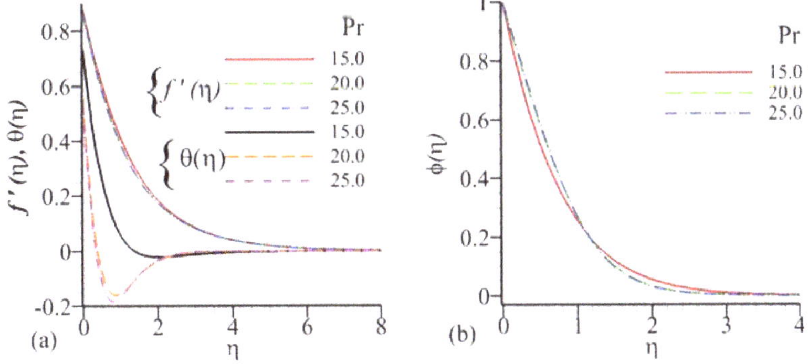

Figure 6. Influence of Pr on velocity profile f', temperature profile θ, and concentration profile ϕ against η.

similar behavior of Figure 6a,b is observed in Figure 7a,b for Dufour parameter as the profiles of velocity, temperature, and Concentration. It is also noticed from Figure 7a,b that the increase in the Soret parameter decreases the radially stretching velocity and temperature profile, while the temperature profile decreases faster than the radial velocity near to the boundary layer. Figure 7b elucidates that the amassed value of the Soret parameter is enhancing in boundary layer thickness with the slip effect parameter effect which is increasing the concentration profile away from the boundary-wall.

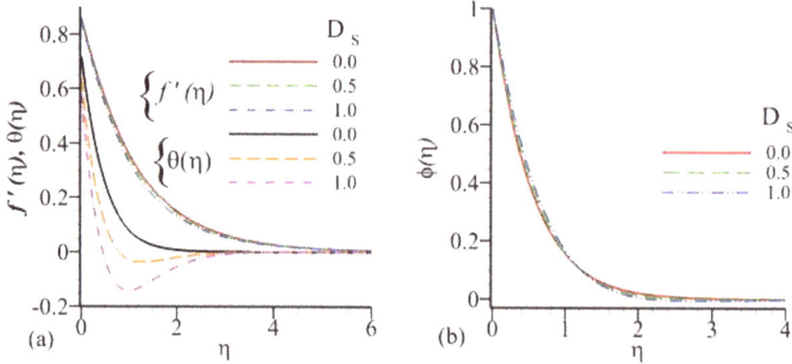

Figure 7. Influence of D_s on velocity profile f', temperature profile θ, and concentration profile ϕ against η.

Figure 8a,b exhibits the influence of Schmidt number Sc on velocity and temperature profiles. Schmidt number is the ratio of momentum and mass diffusivity and is utilized to characterize fluid flows for momentum and mass diffusion convection process. The rise in values of Sc reduces the radially stretching velocity and temperature profiles at a slow rate. The same effect is observed in Figure 8b which shows that increase in Schmidt number has a decreasing effect in concentration profile.

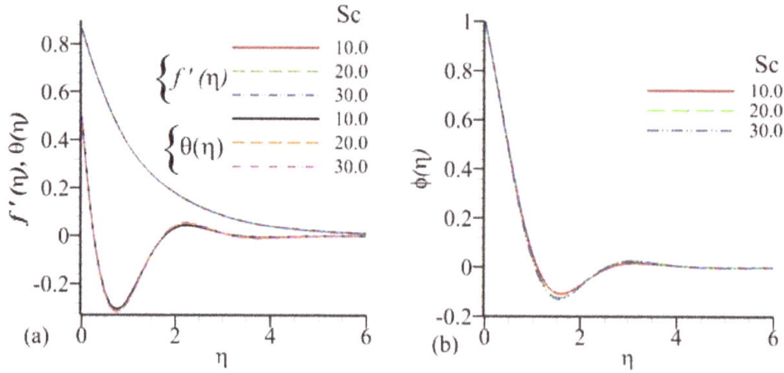

Figure 8. Influence of Sc on velocity profile f', temperature profile θ, and concentration profile ϕ against η.

The Soret effect is where the temperature gradient separates light and heavy molecules. This effect is usually important when there are more than one chemical species in a very large temperature gradient, such as CVD problems and chemical reactors. Figure 9a,b shows the Soret effect of the velocity, temperature, and concentration profiles on a radially stretching sheet. The greater value of Soret parameter results in an increase of velocity, temperature, and concentration profiles away to the boundary wall but closed to the wall little opposite effects can be seen.

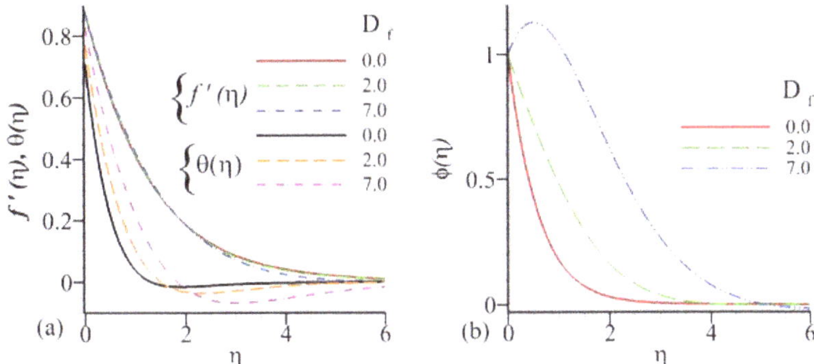

Figure 9. Influence of D_f on velocity profile f', temperature profile θ, and concentration profile ϕ against η.

A chemical reaction is a mechanism leading to the chemical change of one collection of chemicals into another.Figure 10a,b elucidates the impacts of chemical reaction R_0 in the fluid. It is depicted from the graph that increasing values of chemical reaction creates some reaction in the fluid flow and slow down the radially stretching velocity, temperature, and concentration profiles of the fluid. Increasing the temperature profile closed to the wall also show that the boosting the value of chemical reaction that the effect is diminutions the concentration profile away to the wall but closed to the wall it little growing effect find-out.

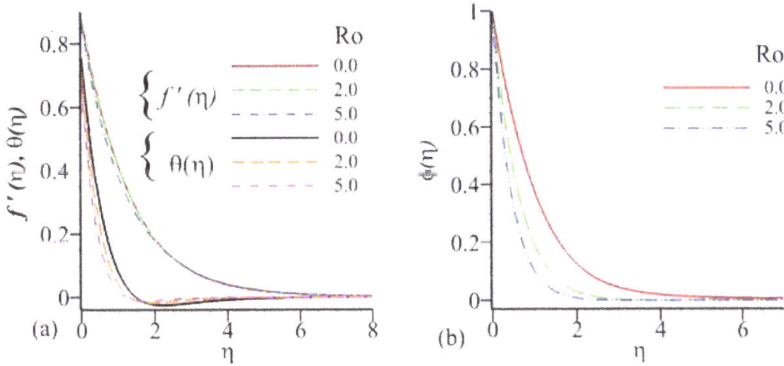

Figure 10. Influence of R_o on velocity profile f', temperature profile θ, and concentration profile ϕ against η.

A parallel crossposting trend is perceived in the velocity, temperature and concentration distribution functions for the increment of Dufour parameter but the opposite behavior is observed in the thermal boundary layer near the boundary. Figures 11 and 12 demonstrate the influence of hydrodynamic and thermal slip on the velocity, temperature, and concentration profiles. It is clear from the Figure 11a that, the consultant radially velocity decreases as the hydrodynamic slip increases but the inverse trend is seen in the temperature and concentration functions (see Figure 11a,b). It is obvious that the velocity, temperature and concentration profiles decrease by increasing the thermal slip value. As the thermal slip parameter value increases, the thermal limit layer thickness decreases even when a small amount of heat is transferred from the surface to the liquid. The velocity, temperature, and concentration boundary layer shrink due to enhancement in thermal slip which is cleary seen in the Figure 12a,b.

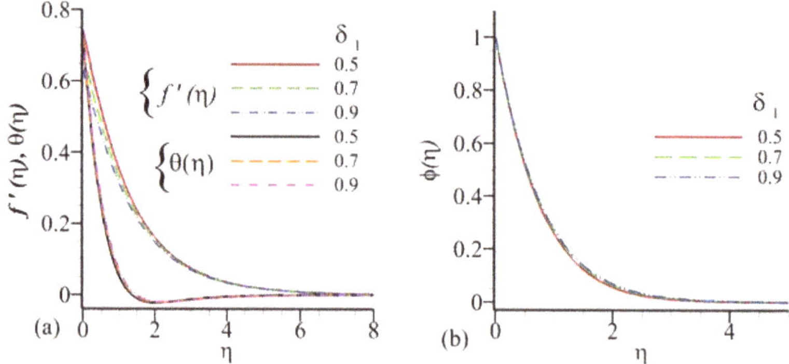

Figure 11. Influence of δ_1 on velocity profile f', temperature profile θ, and concentration profile ϕ against η.

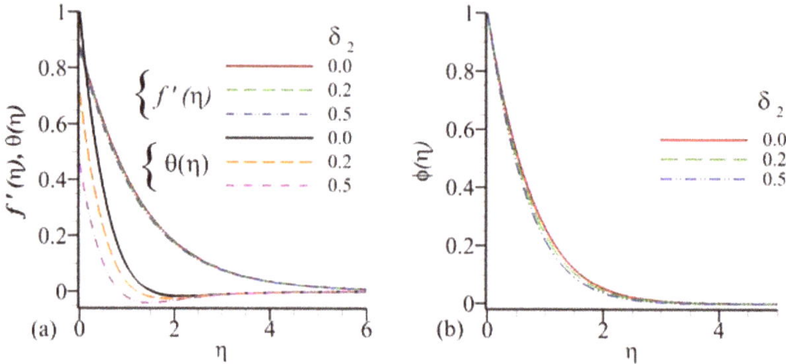

Figure 12. Influence of δ_2 on velocity profile f', temperature profile θ, and concentration profile ϕ against η.

The Figures 13a,b and 14a,b depicts the behavior of suction/injection on fluid velocity, temperature, and concentration profile.The boundary layer of velocity is observed decreasing effect when we enhance in the value of suction parameter $S > 0$ but inverse behavior is seen for velocity, temperature, and concentration profiles when we increase in the values of injection parameter $S < 0$. Figures 15a–c and 16a–c. described the influence of several parameters on the shear stresses, the heat, and mass transfer rates , Figures 15a and 16a illustrate the influence of several parameters on the skin friction factor, that indicate that friction increases with Chemical reaction ,and suction parameter, are enhanced. In results weaker matrix resistance factors to thermo-diffusion flow behaviours in an acceleration leading to increased shearing at the sheet surface and enhance the magnitudes of skin friction factor. Figures 15b and 16b indicates that the variation effect on heat transfer rates for different parameter. The amount of heat transfer rises as, the Chemical reaction ,and suction parameter both are increased.In thermo-diffusion flow, a more inertial impact, obviously aggravates heat diffusion from the radial sheet to the fluid. Figures 15c and 16c show the influence of different parameters on increases in Sherwood number i.e., Chemical reaction ,and suction parameter both are enhancing the magnitude of Sherwood number.

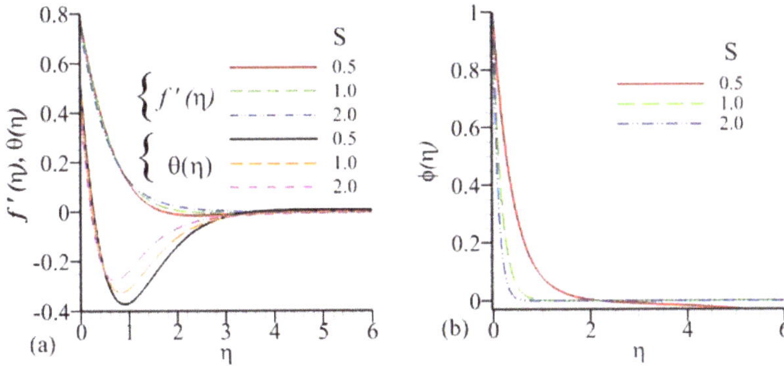

Figure 13. Influence of S on velocity profile f', temperature profile θ, and concentration profile ϕ against η.

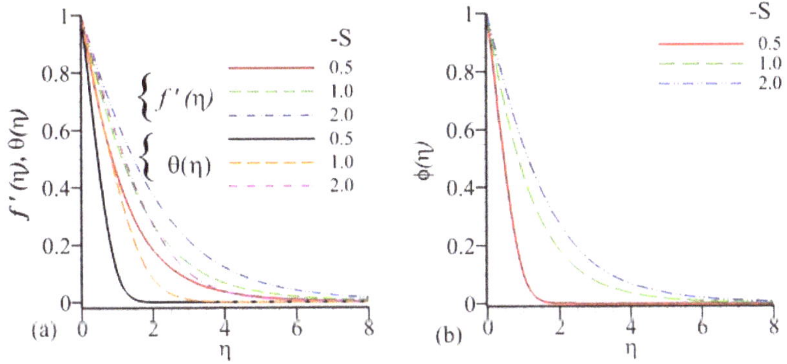

Figure 14. Influence of $-S$ on velocity profile f', temperature profile θ, and concentration profile ϕ against η.

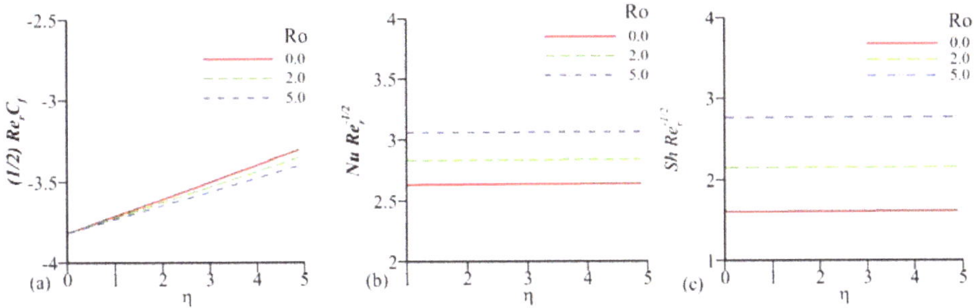

Figure 15. Influence of S on f', θ, and ϕ.

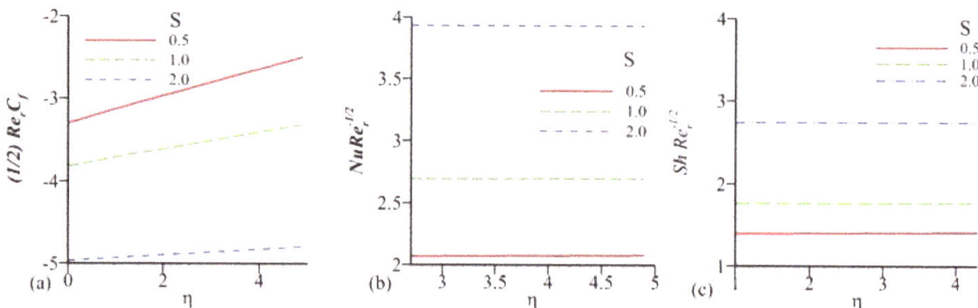

Figure 16. Influence of S on f', θ, and ϕ.

In Table 2 we analysis the variation of physical parameters M, λ, P_r, α, and β, on skin friction co-efficient $-f''(0)$, Nusselt number $-\theta'(0)$, and Sherwood number $-\phi'(0)$.

The following results are concluded from the Table 2. (i) The Skin-friction coefficient is increasing while reducing the local Nusselt and Sherwood numbers through improvement in the increment of Magnetic parameter. (ii) The increment in thermal buoyancy parameter λ, cause decreasing the Skin-friction coefficient while increasing the factor of Nusselt number, and Sherwood number. (iii) With the increasing unsteadiness parameter σ, the Skin-friction , local Nusselt, and Sherwood numbers are also increasing. (iv) The Skin-friction coefficient is increasing with the increment in Prandtl number also increasing the local Nusselt number but opposite trend is observed for Sherwood number. (v) The Skin-friction coefficient and Sherwood number are decreasing with the increasing Casson parameter and increment in the local Nusselt number.

Table 2. Influence of some parameters on $-f''(0)$, $-\theta'(0)$, and $-\phi'(0)$ when $Sc = 10, D_s = D_f = 0.5, R_o = 1.0, \delta_1 = \delta_2 = 0.2, S = 0$.

α	β	M	Pr	λ	$f''(0)$	$-\theta'(0)$	$-\phi'(0)$
0.5	0.3	1.0	10.0	1.0	−2.93927	2.51714	0.97907
1.0					−3.11489	2.65722	1.15877
1.5					−3.27767	2.77597	1.32801
0.5	1.0				−1.87877	2.48897	0.88058
	3.0				−1.48554	2.48759	0.81457
	5.0				−1.40157	2.48426	0.79476
	0.3	2.0			−3.36540	2.50392	0.94527
		4.0			−4.02817	2.48569	0.89369
		6.0			−4.54380	2.47292	0.85671
		1.0	15.0		−2.94933	2.56412	0.88783
			20.0		−2.95470	2.58853	0.84059
			25.0		−2.95780	2.60340	0.81173
			10.0	2.0	−3.08296	3.12406	1.14151
				3.0	−3.08036	3.51020	1.32529
				3.5	−3.07908	3.86004	1.48219

4. Conclusions

The thermo-diffusion and multi-slip effects on an axisymmetric Casson fluid flow over a time-dependent radially stretching sheet with in the presence of chemical reaction is presented. The governing nonlinear PDEs are transformed into a set of highly nonlinear ODEs with the aid of suitable similarity transformations which are solved numerically by utilizing the Keller-Box technique. The computations have been performed for velocity, temperature, and solutal functions for various values of physical parameters. The key conclusions of this work are as follows:

- The velocity profiles are observed to be decreased with increasing values of the Casson, unsteadiness, magnetic, Prandtl number, Dufour, Soret number, chemical, Schmidt number and slips parameters,but the effect of increasing buoyancy parameter values and injection parameters is the opposite.
- Increments in unsteadiness, magnetic field, buoyancy, Prandtl number, Soret, thermal slip, Dufour, and chemical parameters decline the fluid temperature. However, the opposite effect is observed with increasing values of Schmidt number, magnetic, Casson, suction/injection, and hydrodynamic slip parameters.
- The concentration profle are found to be reduced with increasing values of the unsteadiness, buoyancy, Soret, Schmidt number, thermal slip, Prandtl number, chemical reaction, and suction parameters. But the concentration profile are enhanced by increment in the magnetic field, Casson, Prandtl number, Dufour, injection, and hydrodynamic slip parameters.
- The obtained results are presented in graphical and tabular formats. An excellent agreement of our numerical results is obtained with the existing literature which assists with the authenticity of proposed study.
- Destructive chemical reactions are favorable in order to enhance the mass transfer rate.

Author Contributions: F.F. and S.M.I. modeled the problem and wrote the manuscript. B.A. thoroughly checked the mathematical modeling. S.H. checked English corrections. F.F. solved the problem using "Fortran" software. F.F., S.H. and S.M.I. writing—review and editing. F.F., S.M.I. and S.H. contributed to the results and discussions. All authors finalized the manuscript after its internal evaluation.

Funding: This research received no external funding.

Acknowledgments: We are grateful to the Chinese government scholarship council (CSC) for the scholarship award. Also, thankful to Yong Xu for his continued guidance throughout the work.

Conflicts of Interest: The authors declare no conflict of interest.

References

1. Goddard, J.; ACRIVOS, A. An analysis of laminar forced-convection mass transfer with homogeneous chemical reaction. *Q. J. Mech. Appl. Math.* **1967**, *20*, 471–497. [CrossRef]
2. Kong, J.; Cassell, A.M.; Dai, H. Chemical vapor deposition of methane for single-walled carbon nanotubes. *Chem. Phys. Lett.* **1998**, *292*, 567–574. [CrossRef]
3. Kramers, H.A. Brownian motion in a field of force and the diffusion model of chemical reactions. *Physica* **1940**, *7*, 284–304. [CrossRef]
4. Nayak, B.; Mishra, S.; Krishna, G.G. Chemical reaction effect of an axisymmetric flow over radially stretched sheet. *Propuls. Power Res.* **2019**, *8*, 79–84. [CrossRef]
5. Kumar, R.; Kumar, R.; Sheikholeslami, M.; Chamkha, A.J. Irreversibility analysis of the three dimensional flow of carbon nanotubes due to nonlinear thermal radiation and quartic chemical reactions. *J. Mol. Liq.* **2019**, *274*, 379–392. [CrossRef]
6. Bhatti, M.M.; Mishra, S.; Abbas, T.; Rashidi, M.M. A mathematical model of MHD nanofluid flow having gyrotactic microorganisms with thermal radiation and chemical reaction effects. *Neural Comput. Appl.* **2018**, *30*, 1237–1249. [CrossRef]
7. Khan, M.I.; Qayyum, S.; Hayat, T.; Waqas, M.; Khan, M.I.; Alsaedi, A. Entropy generation minimization and binary chemical reaction with Arrhenius activation energy in MHD radiative flow of nanomaterial. *J. Mol. Liq.* **2018**, *259*, 274–283. [CrossRef]
8. Ali, B.; Nie, Y.; Khan, S.A.; Sadiq, M.T.; Tariq, M. Finite Element Simulation of Multiple Slip Effects on MHD Unsteady Maxwell Nanofluid Flow over a Permeable Stretching Sheet with Radiation and Thermo-Diffusion in the Presence of Chemical Reaction. *Processes* **2019**, *7*, 628. [CrossRef]
9. Raza, J. Thermal radiation and slip effects on magnetohydrodynamic (MHD) stagnation point flow of Casson fluid over a convective stretching sheet. *Propuls. Power Res.* **2019**, *8*, 138–146. [CrossRef]
10. Ashraf, M.; Bashir, S. Numerical simulation of MHD stagnation point flow and heat transfer of a micropolar fluid towards a heated shrinking sheet. *Int. J. Numer. Methods Fluids* **2012**, *69*, 384–398. [CrossRef]

11. Daniel, Y.S.; Daniel, S.K. Effects of buoyancy and thermal radiation on MHD flow over a stretching porous sheet using homotopy analysis method. *Alex. Eng. J.* **2015**, *54*, 705–712. [CrossRef]
12. Dhanai, R.; Rana, P.; Kumar, L. Critical values in slip flow and heat transfer analysis of non-Newtonian nanofluid utilizing heat source/sink and variable magnetic field: Multiple solutions. *J. Taiwan Inst. Chem. Eng.* **2016**, *58*, 155–164. [CrossRef]
13. Shahzad, A.; Ali, R.; Hussain, M.; Kamran, M. Unsteady axisymmetric flow and heat transfer over time-dependent radially stretching sheet. *Alex. Eng. J.* **2017**, *56*, 35–41. [CrossRef]
14. Ashraf, M.; Batool, K. MHD flow and heat transfer of a micropolar fluid over a stretchable disk. *J. Theor. Appl. Mech.* **2013**, *51*, 25–38.
15. Shahzad, A.; Ahmed, J.; Khan, M. On heat transfer analysis of axisymmetric flow of viscous fluid over a nonlinear radially stretching sheet. *Alex. Eng. J.* **2016**, *55*, 2423–2429. [CrossRef]
16. Chen, C.H. Laminar mixed convection adjacent to vertical, continuously stretching sheets. *Heat Mass Transf.* **1998**, *33*, 471–476. [CrossRef]
17. Thumma, T.; Beg, O.A.; Kadir, A. Numerical study of heat source/sink effects on dissipative magnetic nanofluid flow from a non-linear inclined stretching/shrinking sheet. *J. Mol. Liq.* **2017**, *232*, 159–173. [CrossRef]
18. Seth, G.; Sarkar, S. MHD natural convection heat and mass transfer flow past a time dependent moving vertical plate with ramped temperature in a rotating medium with Hall effects, radiation and chemical reaction. *J. Mech.* **2015**, *31*, 91–104. [CrossRef]
19. Takhar, H.S.; Agarwal, R.; Bhargava, R.; Jain, S. Mixed convection flow of a micropolar fluid over a stretching sheet. *Heat Mass Transf.* **1998**, *34*, 213–219. [CrossRef]
20. Bhargava, R.; Rana, P. Finite element solution to mixed convection in MHD flow of micropolar fluid along a moving vertical cylinder with variable conductivity. *Int. J. Appl. Math. Mech.* **2011**, *7*, 29–51.
21. Ellahi, R.; Zeeshan, A.; Shehzad, N.; Alamri, S.Z. Structural impact of Kerosene-Al$_2$O$_3$ nanoliquid on MHD Poiseuille flow with variable thermal conductivity: Application of cooling process. *J. Mol. Liq.* **2018**, *264*, 607–615. [CrossRef]
22. Ara, A.; Khan, N.A.; Khan, H.; Sultan, F. Radiation effect on boundary layer flow of an Eyring–Powell fluid over an exponentially shrinking sheet. *Ain Shams Eng. J.* **2014**, *5*, 1337–1342. [CrossRef]
23. Hayat, T.; Waqas, M.; Khan, M.I.; Alsaedi, A.; Shehzad, S. Magnetohydrodynamic flow of burgers fluid with heat source and power law heat flux. *Chin. J. Phys.* **2017**, *55*, 318–330. [CrossRef]
24. Baag, S.; Mishra, S.; Dash, G.; Acharya, M. Numerical investigation on MHD micropolar fluid flow toward a stagnation point on a vertical surface with heat source and chemical reaction. *J. King Saud Univ.-Eng. Sci.* **2017**, *29*, 75–83. [CrossRef]
25. Singh, K.; Kumar, M. Effect of viscous dissipation on double stratified MHD free convection in micropolar fluid flow in porous media with chemical reaction, heat generation and ohmic heating. *Chem. Process Eng. Res.* **2015**, *31*, 75–80.
26. Mabood, F.; Shateyi, S.; Rashidi, M.; Momoniat, E.; Freidoonimehr, N. MHD stagnation point flow heat and mass transfer of nanofluids in porous medium with radiation, viscous dissipation and chemical reaction. *Adv. Powder Technol.* **2016**, *27*, 742–749. [CrossRef]
27. Hayat, T.; Khan, M.I.; Waqas, M.; Alsaedi, A. Mathematical modeling of non-Newtonian fluid with chemical aspects: A new formulation and results by numerical technique. *Colloids Surf. A Physicochem. Eng. Asp.* **2017**, *518*, 263–272. [CrossRef]
28. Mahanthesh, B.; Gireesha, B.; Athira, P. Radiated flow of chemically reacting nanoliquid with an induced magnetic field across a permeable vertical plate. *Results Phys.* **2017**, *7*, 2375–2383. [CrossRef]
29. Mabood, F.; Shateyi, S. Multiple Slip Effects on MHD Unsteady Flow Heat and Mass Transfer Impinging on Permeable Stretching Sheet with Radiation. *Model. Simul. Eng.* **2019**, *2019*. [CrossRef]

Article

Experimental Study on Forced Convection Heat Transfer from Plate-Fin Heat Sinks with Partial Heating

Jae Jun Lee, Hyun Jung Kim and Dong-Kwon Kim *

Department of Mechanical Engineering, Ajou University, Suwon 16499, Korea; sweeperlee@ajou.ac.kr (J.J.L.); hyunkim@ajou.ac.kr (H.J.K.)
* Correspondence: dkim@ajou.ac.kr

Received: 4 September 2019; Accepted: 18 October 2019; Published: 21 October 2019

Abstract: In this study, plate-fin heat sinks with partial heating under forced convection were experimentally investigated. The base temperature profiles of the plate-fin heat sinks were measured for various heating lengths, heating positions, flow rates, and channel widths. From the experimental data, the effects of heating length, heating position, and flow rate on the base temperature profile and the thermal performance were investigated. Finally, the characteristics of the optimal heating position were investigated. As a result, it was shown that the optimal heating position was on the upstream side in the case of the heat sinks under laminar developing flow, as opposed to the heat sinks under turbulent flow. It was also shown that the optimal heating position could change significantly due to heat losses through the front and back of the heat sink, while the effects of the heat loss through the sides of the heat sink on the optimal heating position were negligible. In addition, it was shown that the one-dimensional numerical model with empirical coefficients could predict the important trends in the measured temperature profiles, thermal resistances, and optimal heating lengths.

Keywords: plate-fin heat sink; partial heating; forced convection

1. Introduction

The demand for high-performance and compact products has resulted in a continuous increase in heat dissipation from electronic devices in various systems, such as power converters, supercomputers, and electric vehicles [1,2]. The high heat dissipation from electronic products results in a high junction temperature, which negatively affects the overall performance and durability of the product. Therefore, efficient cooling of these devices is essential and various cooling methods have been developed. In addition, the efficient cooling is also essential for various thermal systems such as transcritical CO_2 refrigeration systems [3], ice storage tanks [4], wind tunnels [5], and heat exchanger/reactors [6]. The most common cooling method is the use of plate-fin heat sinks under forced convection. Therefore, many previous studies have focused on plate-fin heat sinks [7–11]. The thermal performance of these heat sinks is best when the heated base area is equal to the total base area, because the spreading resistance is minimized [12]. However, in many cases, the total base area of the heat sink is larger than the heated base area, as shown in Figure 1. It is because the amount of heat dissipation increases as the total base area increases, and the large total base area is required to sufficiently dissipate heat from high-performance and compact electronic devices, in many cases. For example, small Insulated Gate Bipolar Transistors (IGBTs) are generally mounted on large heat sinks in power electronics [13].

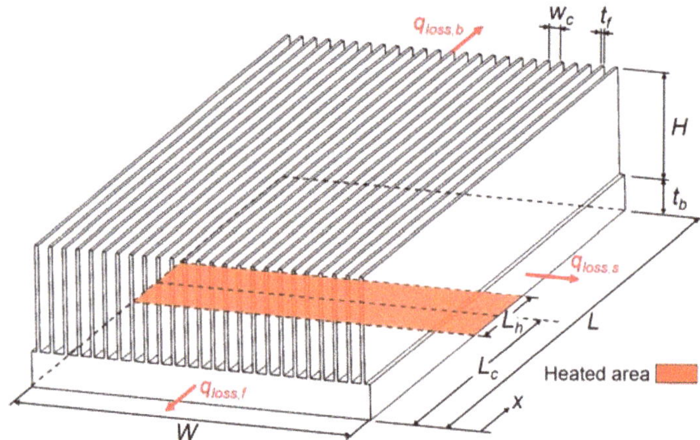

Figure 1. Plate-fin heat sinks with partial heating.

Many researchers have investigated the thermal performance of plate-fin heat sinks with partial heating under forced convection. These research works are well-summarized in Yoon et al. [14]. For example, Culham et al. investigated the role of spreading resistance of the heat sink base caused by partial heating in the design and the selection of the plate-fin heat sinks [15]. They showed that the spreading resistance could be significant without proper thermal design of the heat sinks. Ellison analytically solved a three-dimensional heat conduction equation for obtaining temperature fields in the heat sink bases and for calculating spreading resistances for various heat source sizes [16]. He showed that his results could be used for designing plate-fin heat sinks. Cho et al. experimentally investigated the effects of non-uniform mass flow distributions and heat flux conditions on the thermal performance of microchannel heat sinks [17]. They observed that the microchannel heat sink should be designed by considering the heat flux conditions. Yoon et al. also experimentally investigated the thermal performance of plate-fin and strip-fin minichannel heat sinks, under partial heating [18]. Lelea analyzed the effects of the heating position on the thermal performance of a microchannel heat sink by numerically solving the conjugate heat transfer problem [19]. He observed that the heating position influences the thermal characteristics and that upstream heating has a better thermal performance than central or downstream heating. Toh et al. also numerically calculated the three-dimensional velocity and temperature fields in a partially heated microchannel heat sink by using a finite volume method [20]. Although all of these studies are systematic and provide useful results, they used water or refrigerant as the working fluid and focused on minichannel heat sinks or microchannel heat sinks. As a result, the information which these studies provide for macroscale heat sinks with air as the working fluid is limited. However, macroscale heat sinks with air as the working fluid are commonly used in various systems. Therefore, there have been several studies for investigating the thermal performance of macroscale plate-fin heat sinks under partial heating with air as the working fluid. For example, Emekwuru et al. numerically investigated the influence of three different partial heating positions on the thermal performance of air-cooled macroscale heat sinks [21]. They showed that the thermal resistance is minimized when the partial heating position is centrally located, for most Reynolds numbers. Yoon et al. extensively conducted numerical studies on the thermal performance of plate-fin heat sinks according to the partial heating position to determine the optimal heating position [14]. They showed that the thermal performance is maximized when the heating position is located on the downstream side near the center of the heat sink.

However, first, to the best of our knowledge, intensive experimental studies on the thermal performances of macroscale plate-fin heat sinks under partial heating with air as the working fluid have not been systematically conducted for various heating locations, flow rates, and channel widths.

In particular, previous studies on macroscale plate-fin heat sinks [14,21] was focused on heat sinks under turbulent flows with very high flow rates, and it is uncertain whether their conclusions are applicable for heat sinks under laminar developing flows with low flow rates.

Second, heat sinks used in the practical situation showed heat losses through the front, back, and sides of the heat sink ($q_{loss,f}$, $q_{loss,b}$, and $q_{loss,s}$ in Figure 1). However, there were no previous studies on the effects of these heat losses on the optimal heating position of macroscale plate-fin heat sinks under partial heating.

Third, a full three-dimensional numerical simulation could estimate the thermal performance accurately but required a time-consuming code writing process. In addition, commercial numerical simulation programs might have cost issues. Therefore, sometimes it is useful to perform several experiments and to investigate the effects of important engineering parameters on thermal performance by using a one-dimensional model that can predict experimental results.

In this study, plate-fin heat sinks under partial heating with air as the working fluid were experimentally investigated. The base temperature profiles of plate-fin heat sinks under laminar developing flows were measured for various heating lengths, heating positions, flow rates, and channel widths. From the experimental data, the effects of various engineering parameters on the base temperature profile and the thermal performance were investigated. Finally, the characteristics of the optimal heating position were investigated. It is shown that the conclusions of the previous studies [14,21] on optimal heating position of plate-fin heat sink under turbulent flow did not apply to the heat sink under the laminar developing flow examined in this study. In is also shown that the optimal heating position could change significantly due to heat losses through the front and back of the heat sink. In addition, a simple one-dimensional numerical model for predicting the thermal performances is presented.

2. Experimental Procedure

A schematic diagram of a heat sink is shown in Figure 1. The length L, width W, height H, base thickness t_b, and fin thickness t_f of the heat sink were 20 cm, 10 cm, 3 cm, 1 cm, and 1 mm, respectively. The experiment was conducted for five different combinations of heating lengths and heating positions, four different flow rates, and two different channel widths, as listed in Table 1. In this table, N, w_c, L_h, and L_c are the fin number, channel width, heated length, and the distance from the upstream end to the center of the heated area, respectively. The heat sink was made of SUS304 ($k = 16.2$ W/m K). As shown in Figure 2, 22 T-type thermocouples were attached to the base of the heat sink. More specifically, in order to measure the profile of the base temperature T_b along the flow direction, thermocouples were installed at 11 points in the x direction. The thermocouples were attached at two points in the y direction to check that the measured base temperature profile along the flow direction was reliable. Silver epoxy was used to minimize contact thermal resistance between the thermocouple and the base. Four other T-type thermocouples were placed at the inlet of the test section to measure the ambient temperature T_{amb}. As shown in Figure 3, the signals from the thermocouples were converted to temperature data using a data acquisition system (34970A DAQ; Agilent Technology) and were stored in a computer. As shown in Figure 2, five film heaters of the same size were attached to the bottom surface of the heat sink base to locally apply heat to the heat sink. The film heater was made of SUS304 and was attached using a Kapton tape. These five heaters were powered and controlled by a DC power supply (PL-3005D; Protek) to be turned on and off, independently.

The total power input to the heater was fixed at 40 W. Five different combinations of heating lengths and heating positions were investigated in this study, as listed in Table 1. As shown in Figure 3, the heat sink with the attached thermocouples and heaters was installed inside a rectangular duct in acrylic. The duct was connected to a wind tunnel system with a nozzle flow meter and a blower (TB-150; Innotech) to measure the flow rate of air while generating air flow in the duct. The nozzle flow meter consisted of the nozzle and the differential pressure meter (FCO332; Furness Controls), and it measured the volume flow rate by measuring pressure difference between the inlet and the outlet of the nozzle caused by flow restriction by the nozzle. Four different flow rates were investigated in this study, as listed in Table 1. Each measurement was repeated three times to ensure the reliability of data. The temperature was measured until the temperature change dropped to less than 0.1 °C for 10 min. After the experiment, the uncertainty of the data were analyzed, and details of the analysis were described in Reference [22] and Appendix A.

Table 1. Design parameters for the experiment.

N	w_c [mm]	L_h/L	L_c/L	Flow Rate Q [m³/min]
25	3	1.0	0.5	0.2, 0.4, 0.6, 0.8
		0.6	0.5	0.2, 0.4, 0.6, 0.8
		0.2	0.3	0.2, 0.4, 0.6, 0.8
			0.5	0.2, 0.4, 0.6, 0.8
			0.7	0.2, 0.4, 0.6, 0.8
13	7	1.0	0.5	0.2, 0.4, 0.6, 0.8
		0.6	0.5	0.2, 0.4, 0.6, 0.8
		0.2	0.3	0.2, 0.4, 0.6, 0.8
			0.5	0.2, 0.4, 0.6, 0.8
			0.7	0.2, 0.4, 0.6, 0.8

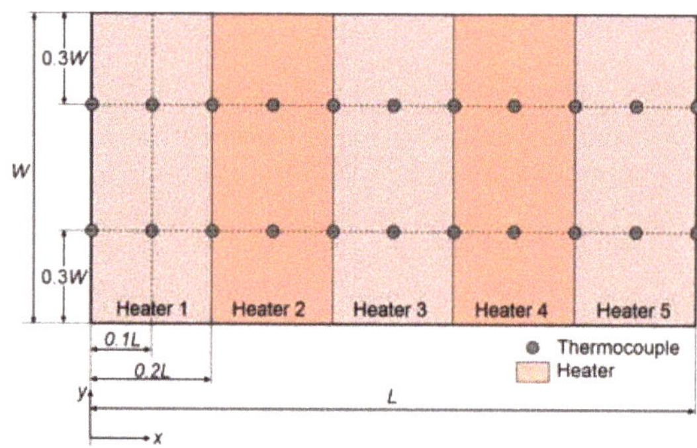

Figure 2. Positions of thermocouples and heaters on the heat sink base.

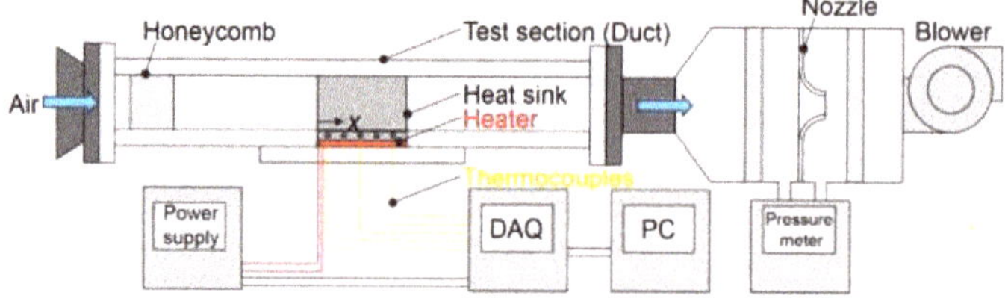

Figure 3. Schematic diagram of the experimental setup.

3. Results

Figure 4a–e show the profiles of the difference between the base temperature T_b and the ambient temperature T_{amb}, along the flow direction for five different combinations of heating lengths and heating positions and four different flow rates when the channel width was 7 mm. The various trends of the temperature profile observed in these figures were as follows.

(a)   In all cases shown in Figure 4a–e, regardless of the value of x, the base temperature decreased as the flow rate increased. This was mainly because the temperature of the fluid flowing inside the heat sink decreased as the flow rate increased. This was also because the local heat transfer coefficient, which was inversely proportional to the temperature difference between the fluid temperature and the base temperature, increased as the flow rate increased.

(b)   The extent to which the base temperature decreased with an increase in the flow rate, decreased as the flow rate increased.

(c)   The point at which the base temperature was maximized always existed in the portion where the heat sink was heated. When the size of the heating portion was equal to the size of the heat sink, the point at which the base temperature was maximized lay almost at the end of the heating portion where x was the largest. When the size of the heating portion was much smaller than the size of the heat sink, the point where the base temperature was maximized was almost at the center of the heating portion.

(d)   The base temperature at $x = L_c + \Delta x$ was larger than the base temperature at $x = L_c - \Delta x$, regardless of the value of Δx. This was mainly because the fluid temperature monotonically increased as x increased.

(e)   As shown in Figure 4a, when the heat sink was uniformly heated, the base temperature increased monotonically with an increase in x. This was because the temperature of the fluid increased as x increased. This was also because the heat transfer coefficient, which was inversely proportional to the temperature difference between the fluid temperature and the base temperature, decreased as x increased.

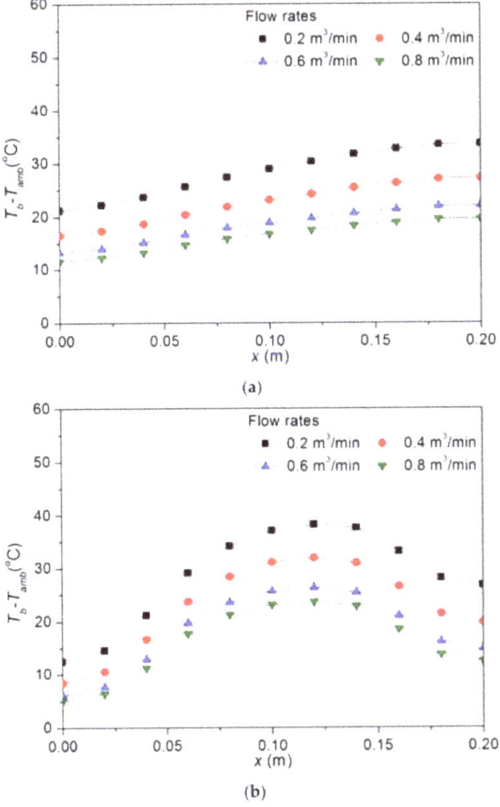

(a)

(b)

Figure 4. *Cont.*

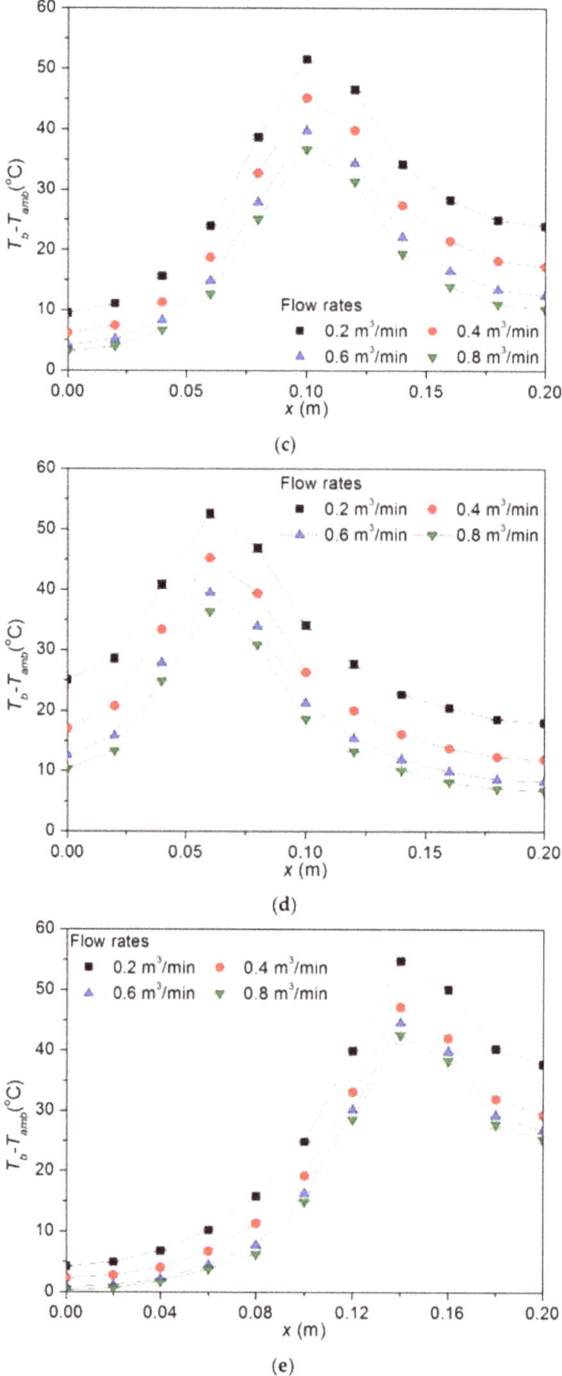

Figure 4. Profiles of difference between base temperature and ambient temperature (w_c = 7 mm). (**a**) L_h/L = 1.0, L_c/L = 0.5; (**b**) L_h/L = 0.6, L_c/L = 0.5; (**c**) L_h/L = 0.2, L_c/L = 0.5; (**d**) L_h/L = 0.2, L_c/L = 0.3; and (**e**) L_h/L = 0.2, L_c/L = 0.7.

Figure 5a–e show the profiles of the difference between the base temperature T_b and the ambient temperature T_{amb} along the flow direction when the channel width was 3 mm. The trends in the temperature profile were consistent with the previously described trends for the channel width of 7 mm.

Figure 6a,b show the thermal resistances for five different combinations of heating lengths and heating positions, four different flow rates, and two different channel widths. Here, the thermal resistance R was defined as the difference between the maximum temperature of the heat sink base $T_{b,max}$ and the ambient temperature T_{amb} divided by the amount of heat q applied to the heat sink. Therefore,

$$R = \frac{T_{b,max} - T_{amb}}{q}. \tag{1}$$

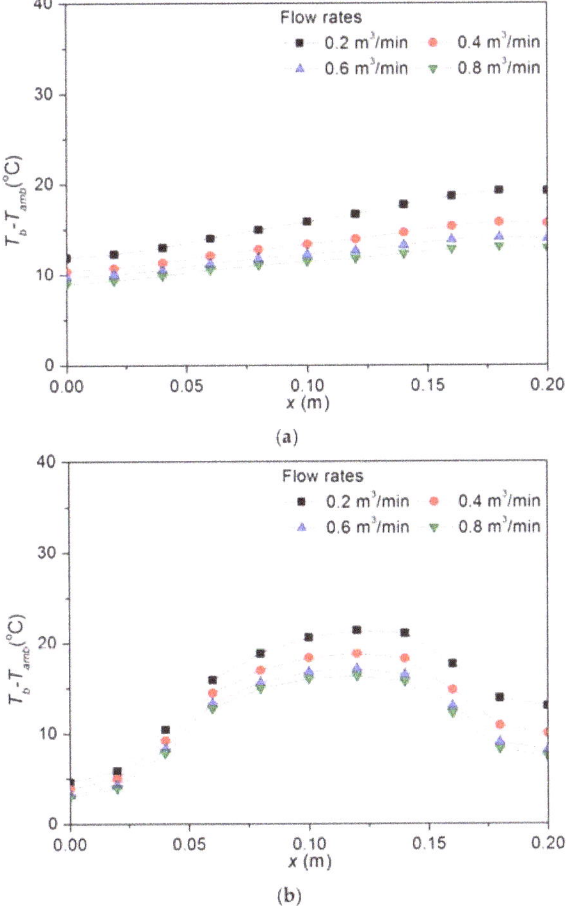

(a)

(b)

Figure 5. *Cont.*

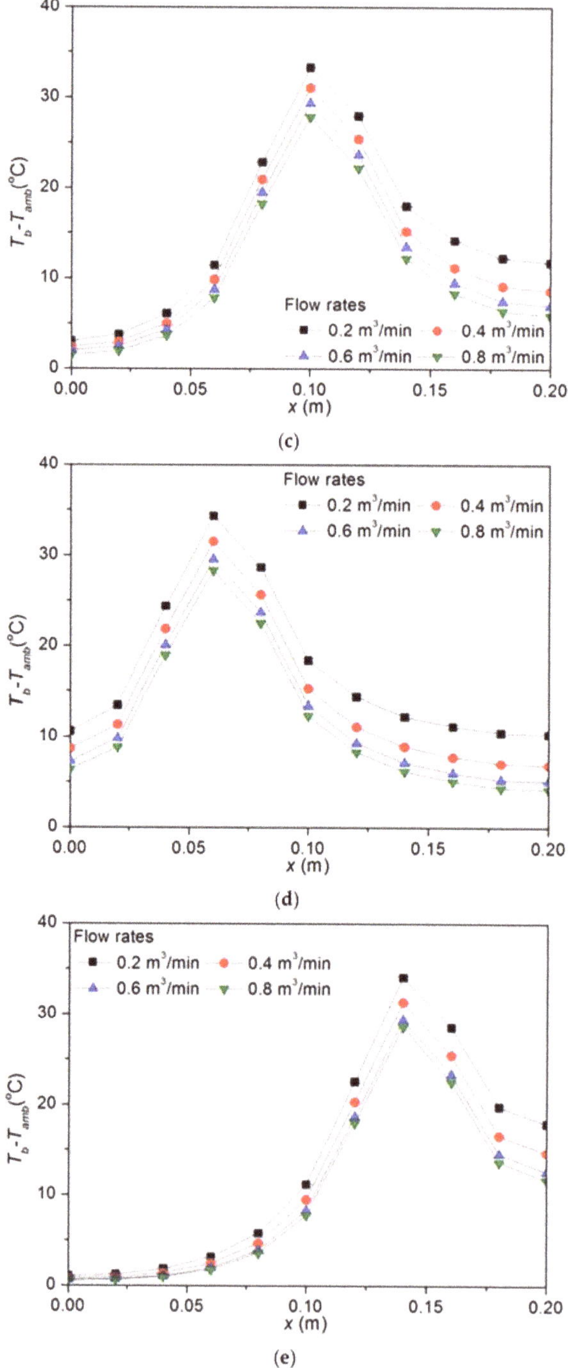

Figure 5. Profiles of difference between base temperature and ambient temperature (w_c = 3 mm). (**a**) L_h/L = 1.0, L_c/L = 0.5; (**b**) L_h/L = 0.6, L_c/L = 0.5; (**c**) L_h/L = 0.2, L_c/L = 0.5; (**d**) L_h/L = 0.2, L_c/L = 0.3; and (**e**) L_h/L = 0.2, L_c/L = 0.7.

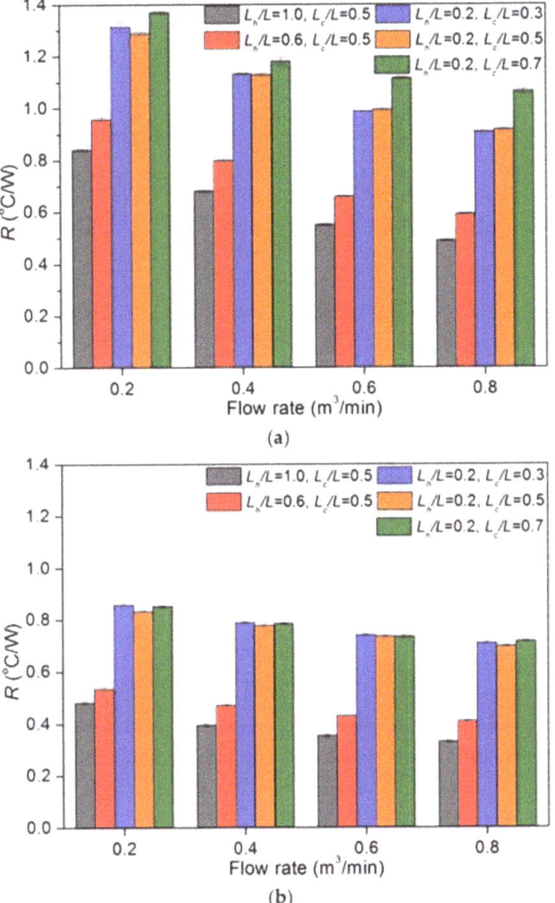

Figure 6. Thermal resistances. (**a**) $w_c = 7$ mm; and (**b**) $w_c = 3$ mm.

The various characteristics of thermal resistance observed in these figures are as follows.

(a) Regardless of the heating position and channel width, the thermal resistance decreased monotonically as the flow rate increased. This was because the temperature of the cooling fluid filling the inside of the heat sink decreased as the flow rate increased.

(b) Regardless of the channel width and flow rate, the thermal resistance increased as the heating length decreased. This was because the heat sink base was more heavily heated locally, owing to an increase in heat flux (amount of heat per unit area) applied to the heated portion of the heat sink base as the heating length decreased. When the heat sink base was uniformly heated, heat was uniformly dissipated from the heat sink to the fluid through the heat sink surface. In contrast, only a part of the surface was mainly used for the heat dissipation in the case of the locally heated heat sink.

(c) There was an optimal heating position that minimized the thermal resistance when the heating length was fixed. However, as shown in Figure 6a, the optimal heating position was not fixed and could be changed by changing the flow rate.

Finally, to investigate the characteristics of characteristics of the optimal heating position, we developed a relatively simple one-dimensional numerical model that could roughly predict the experimental results (further explanation on a one-dimensional numerical model is given in Appendix B). Finally, Figure 7 shows the thermal resistances for various heating positions and flow rates when the heating length was fixed. The characteristics of the optimal heating position observed from Figure 7 were as follows.

(a) When the flow rate was fixed, there was an optimal heating position that minimized the thermal resistance.

(b) The optimal heating position was not independent of the flow rate. In this study, the optimal heating position moved toward the smaller x as the flow rate increased. Therefore, when controlling the flow rate from the fan in an actual situation, this fact should be considered to determine the heating position.

(c) If the heating position moved slightly in the positive x direction or in the negative x direction from the optimal heating position, the thermal resistance did not increase significantly. Therefore, in an actual situation, a slight change of the heating position from the optimal heating position was acceptable.

(d) However, if the heating position was far from the optimal heating position, the thermal resistance could be very large. Therefore, in an actual situation, the heating position should not be close to the front or rear ends of the heat sink.

(e) As presented earlier, Emekwuru et al. showed that the thermal resistance was minimized when the heating position was centrally located for most Reynolds numbers [21]. On the other hand, Yoon et al. showed that the thermal performance was maximized when the heating position was located on the downstream side near the center of the heat sink [14]. However, these conclusions from the previous studies did not apply to the situation examined in this study. Figure 7 indicates that the thermal performance was maximized when the heating position was located on the upstream side. In addition, Figure 6a also shows that the optimal heating position was located on the upstream side in some cases. The main reason why the conclusions from the previous studies did not apply to the situation examined in this study was that the previous study was focused on heat sinks under turbulent flows with very high flow rates and the present experiments were conducted for heat sinks under laminar developing flows. The fact that Lelea also observed that upstream heating had a better thermal performance than central or downstream heating for the heat sink under laminar flow in his study [19] supported this reasoning, although his study was focused on water-cooled microchannel heat sinks.

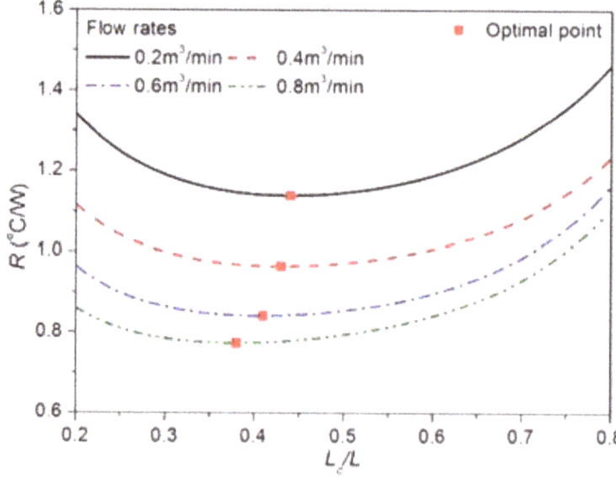

Figure 7. Thermal resistances for various heating positions and flow rates (w_c = 7 mm, L_h/L = 0.2).

The heat sinks used in the practical situations showed heat losses through the front, back, and sides of the heat sink ($q_{loss,f}$, $q_{loss,b}$, and $q_{loss,s}$ in Figure 1). Therefore, in this study, effects of these heat losses on the optimal heating position were investigated. Figures 8–10 show the thermal resistances of the heat sinks calculated by one-dimensional models for various thermal resistances for the heat losses through the front, back, and sides of the heat sinks, respectively. In these figures, $R_{loss,f}$ and $R_{loss,b}$ are the thermal-resistances-related heat losses through the front and back of the heat sink, respectively. $R'_{loss,s}$ is the thermal resistance per unit length related to heat loss through the sides of the heat sink, as explained in Appendix B. The characteristics of the optimal heating position observed in Figures 8–10 were as follows.

(a) The optimal heating position could change significantly due to heat losses through the front and back of heat sink. Therefore, even when the heat sink with the same dimensions was under the same flow rate, the optimal heating position could be completely different depending on how the heat sink was connected to the other devices in the thermal system.

(b) The optimal heating position moved toward the smaller x as the heat loss through the front of the heat sink increased.

(c) The optimal heating position moved toward the larger x as the heat loss through the back of the heat sink increased.

(d) The optimal heating position was almost independent to the heat loss through the sides of the heat sink.

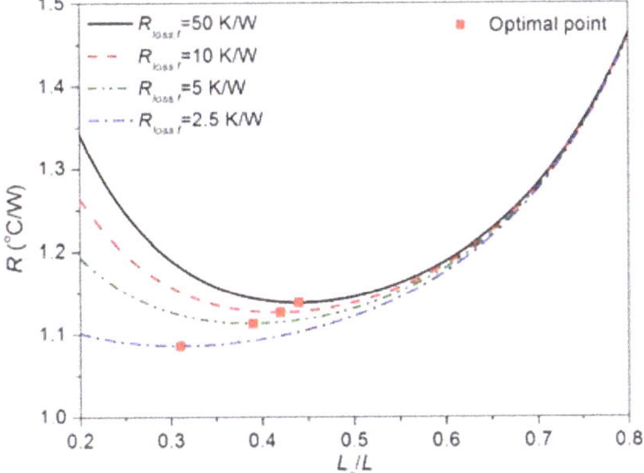

Figure 8. Thermal resistances for various heat losses through the front of the heat sink ($Q = 0.2$ m³/min, $w_c = 7$ mm, $L_h/L = 0.2$).

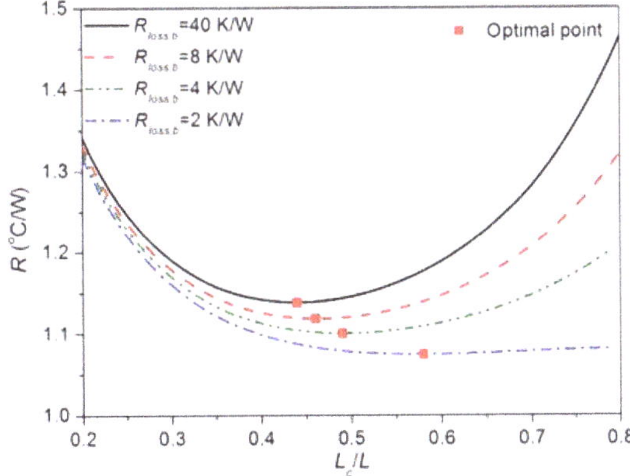

Figure 9. Thermal resistances for various heat losses through the back of the heat sink ($Q = 0.2$ m³/min, $w_c = 7$ mm, $L_h/L = 0.2$).

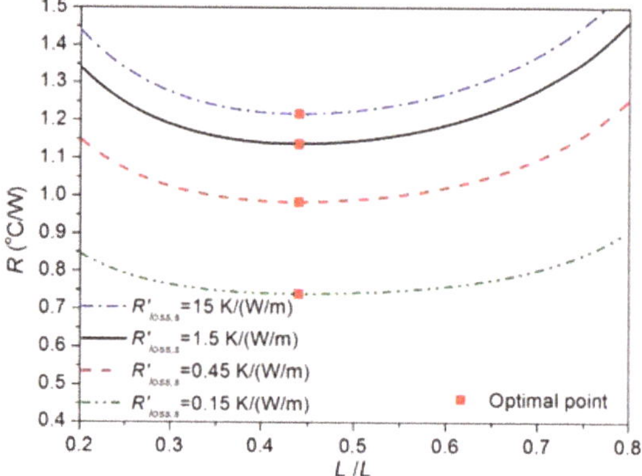

Figure 10. Thermal resistances for various heat losses through the sides of the heat sink ($Q = 0.2$ m^3/min, $w_c = 7$ mm, $L_h/L = 0.2$).

In addition, Figure 11 shows the thermal resistances for various heated lengths. As shown in the figure, the optimal heating position also did not depend strongly on the heated length, if the change in heat transfer coefficient due to the heated length was not significant.

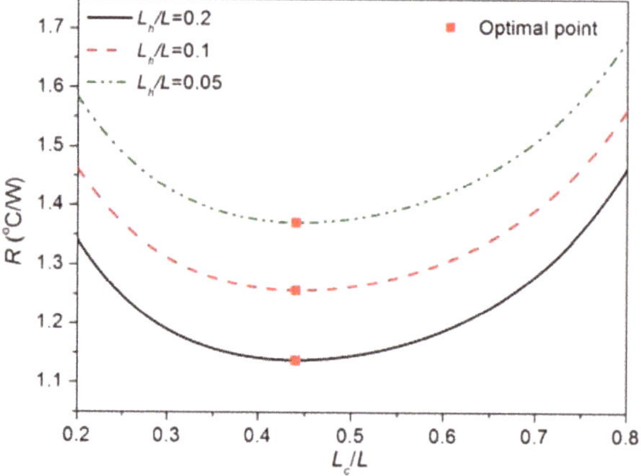

Figure 11. Thermal resistances for various heated lengths ($Q = 0.2$ m^3/min, $w_c = 7$ mm).

4. Conclusions

In this study, macroscale plate-fin heat sinks under partial heating with air as the working fluid were experimentally investigated. The base temperature profiles of the plate-fin heat sinks under laminar developing flows were measured for various heating lengths, heating positions, flow rates, and channel widths. From the experimental data, the effects of various engineering parameters on the base temperature profile and the thermal performance were investigated. Finally, the characteristics of the optimal heating position were investigated.

The following were the new findings on the optimal heating position of the partially heated plate-fin heat sinks:

(a) The optimal heating position was on the upstream side in the case of the heat sinks under a laminar developing flow that were investigated in the present study. On the contrary, as shown in a previous study [14], the optimal heating position was on the downstream side in the case of the heat sinks under turbulent flow.

(b)   The optimal heating position could change significantly due to heat losses through the front and back of heat sink. Therefore, even when the heat sink with the same dimensions was under the same flow rate, the optimal heating position could be completely different depending on how the heat sink was connected to other devices in the thermal system.

(c)   The optimal heating position was independent to the heat loss through the sides of the heat sink. The optimal heating position also did not depend strongly on the heated length.

(d)   The one-dimensional numerical model with empirical coefficients could reflect the important trends in the measured temperature profiles, thermal resistances, and optimal heating lengths.

In addition, the following are some characteristics of the partially heated plate-fin heat sinks presented in this study:

(a)   The point at which the base temperature was maximized always existed in the portion where the heat sink was heated. When the size of the heating portion was equal to the size of the heat sink, the point at which the base temperature was maximized lay almost at the end of the heating portion where x was the largest. When the size of the heating portion was much smaller than the size of the heat sink, the point where the base temperature was maximized was almost at the center of the heating portion.

(b)   Regardless of the heating position and the channel width, the thermal resistance decreased monotonically as the flow rate increased. Regardless of the channel width and flow rate, the thermal resistance increased as the heating length decreased.

(c)   The optimal heating position was not independent of the flow rate.

(d)   If the heating position moved slightly in the positive x direction or in the negative x direction from the optimal heating position, the thermal resistance did not increase significantly. However, if the heating position was far from the optimal heating position, the thermal resistance could be very large.

Author Contributions: J.J.L. and D.-K.K. conducted the investigation; D.-K.K. and H.J.K. performed the formal analyses; D.-K.K. wrote the manuscript.

Funding: This research was supported by Nano-Material Technology Development Program through the National Research Foundation of Korea (NRF) funded by Ministry of Science and ICT (NRF 2011 0030285).

Conflicts of Interest: The authors declare no conflict of interest. The funders had no role in the design of the study; in the collection, analyses, or interpretation of data; in the writing of the manuscript, or in the decision to publish the results.

Nomenclature

A_s	solid cross-sectional area
C_1, C_2, C_3, C_4	empirical coefficients
c_f	fluid heat capacity
H	height of heat sink
h	heat transfer coefficient
k_s	solid thermal conductivity
L	length of heat sink
L_c	distance from the upstream end to the center of the heated area
L_h	heated length
$\dot{m}$	mass flow rate
N	fin number

P	wetted perimeter
Q	flow rate
q	amount of heat applied to heat sink
Q'	heat input per unit length
$q_{loss,b}$	heat loss through the back of the heat sink
$q_{loss,f}$	heat loss through the front of the heat sink
$q_{loss,s}$	heat loss through the sides of the heat sink
R	thermal resistance
$R_{loss,b}$	thermal resistance related heat loss through the back of the heat sink
$R_{loss,f}$	thermal resistance related heat loss through the front of the heat sink
$R'_{loss,s}$	thermal resistance per unit length related to heat loss through the sides of the heat sink
T_{amb}	ambient temperature
T_b	base temperature
$T_{b,max}$	maximum temperature of the heat sink base
t_b	base thickness
t_f	fin thickness
W	width of heat sink
w_c	channel width
η_o	overall surface efficiency

Appendix A.

Experimental Uncertainties

According to the manufacturers' calibration results, the instrument bias errors in the measurement are as shown in Table A1.

Table A1. Instrument bias errors.

Measurement Type	Error
Temperature	<0.5 K
Voltage	<0.01%
Current	<0.1%
Pressure	<0.5%

Appendix B.

Detailed Explanation for the One-Dimensional Numerical Model

This appendix describes how to fit the thermal resistance as a function of flow rate and heating position by using a one-dimensional numerical model in this study.

The easiest method is to directly fit the thermal resistance shown in Figure 6 as a function of the heating position and flow rate by using some suitable functional form. However, if this method is used, the important trends in the measured temperature profiles shown in Figures 4 and 5 are ignored and not reflected, so there is a high probability that the results would not be engineeringly meaningful and would be less reliable.

Therefore, in this study, we developed a relatively simple one-dimensional numerical model with empirical coefficients that could roughly predict the temperature profiles. Then empirical coefficients were obtained by determining the values that minimized the root mean square error (RMSE) between the temperatures obtained through the experiment and the temperatures obtained from the numerical model. Finally, the thermal resistance was calculated as a function of the heating position and the flow rate from the model; presented in Figure 7.

Further explanation on the model is given as follows. According to References [23] and [24], the fluid temperature T_f and the base temperature T_b averaged in the direction perpendicular to the flow are given by the following Equations (A1) and (A2), respectively.

$$k_s A_s \frac{d^2 T_b}{dx^2} + \eta_o h P(T_f - T_b) - \frac{1}{R'_{loss,s}}(T_b - T_{amb}) + q' = 0 \tag{A1}$$

$$- \dot{m} c_f \frac{d T_f}{dx} + \eta_o h P(T_b - T_f) = 0 \tag{A2}$$

Here, k_s, A_s, η_o, h, P, $R'_{loss,s}$, q', $\dot{m}$, and c_f are the solid thermal conductivity, solid cross-sectional area, overall surface efficiency [25], heat transfer coefficient, wetted perimeter, thermal resistance per unit length related to heat loss through the sides of the heat sink ($q_{loss,s}$ in Figure 1), heat input per unit length, mass flow rate, and fluid heat capacity, respectively. The first and second terms in Equation (A1) are related to the conduction in the heat sink in the flow direction and the convection from the heat sink to the fluid, respectively. The third term is related to the heat loss from the heat sink to the surroundings in the direction perpendicular to the flow, and the fourth term is related to the heat input from the heater to the heat sink. The first and second terms in Equation (A2) are related to the fluid enthalpy change in the flow direction and the convection from the heat sink to the fluid, respectively. The governing equations were discretized by the control-volume-based finite difference method, and discretization equations were calculated using the Gauss–Seidel method.

The boundary conditions for these equations are given as follows.

$$T_f(x = 0) = T_{amb} \tag{A3}$$

$$k_s A_s \frac{dT_b}{dx}\bigg|_{x=0} = \frac{1}{R_{loss,f}}(T_b(x = 0) - T_{amb}), \; k_s A_s \frac{dT_b}{dx}\bigg|_{x=L} = -\frac{1}{R_{loss,b}}(T_b(x = L) - T_{amb}) \tag{A4}$$

The first boundary condition (Equation (A3)) indicated that the fluid temperature was at the ambient temperature at the inlet. The second boundary condition (Equation (A4)) indicated that there were heat losses in the -x direction from the upstream end face and in the +x direction from the downstream end face of the heat sink. In Equation (A4), $R_{loss,f}$ and $R_{loss,b}$ are the thermal resistances related heat losses through the front and back of the heat sink ($q_{loss,f}$ and $q_{loss,b}$, in Figure 1), respectively. To obtain the temperature profiles from the numerical model, we must know the value of the local heat transfer coefficient h. However, the exact value of the local heat transfer coefficient could be obtained only from a three-dimensional numerical analysis or from a precise bulk-mean fluid temperature measurement. However, such a complex numerical analyses or experiment was beyond the scope of this study, because the reason for using the numerical model in this study was to predict the tendency of the profile of temperatures qualitatively, using a simple method. Therefore, in this study, the local heat transfer coefficient was assumed to be as follows.

$$h = C_1(1 + C_2 L_c/L) - C_3(x/L)(1 + C_4 L_c/L) \tag{A5}$$

This equation is not perfect, but it reflects the following important facts. (a) As x increases, the local heat transfer coefficient tends to decrease. (b) The local heat transfer coefficient changes owing to the change in the heating position.

In this study, empirical coefficients were obtained by determining the values that minimized the root mean square error (RMSE) between the temperatures obtained through the experiment and the temperatures obtained from the numerical model.

$$RMSE = \sum_{i=1}^{11} \sqrt{T_{b,measured}(x_i) - T_{b,calculated}(x_i)} \tag{A6}$$

In Equation (A6), $T_{b,measured}$, $T_{b,calculated}$, and x_i are the measured base temperature, base temperature calculated from the model, and x coordinates of the measurements points, respectively. More specifically, the values of $R_{loss,f}$, $R_{loss,b}$, and $R_{loss,s}'$, which are not directly related to convective heat transfer, are obtained first by determining the values that minimize the RMSE for a uniformly heated heat sink without fluid flow, as shown in Figure A1. Subsequently, the remaining terms C_1, C_2, C_3, and C_4, which are related to convective heat transfer, are obtained by determining the values that minimize the RMSE for a partially heated heat sink with fluid flow, as shown in Figure A2 for example.

Figure A2 compares the profiles of the temperature difference between the base temperature and the ambient temperature calculated using the numerical model with the experimental results for several cases with the same heating length and different heating positions. It could be observed that the base temperature profiles were qualitatively well-predicted by the model, although the simple numerical model could not perfectly predict the temperatures for all positions.

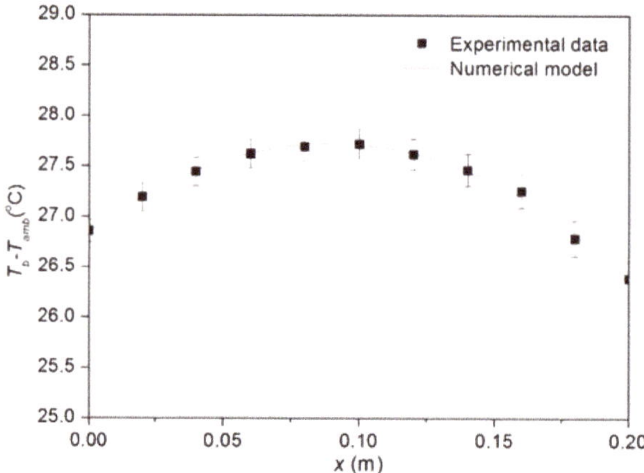

Figure A1. Comparison of temperature differences from experimental data and results of the numerical model for L_h/L of 1, channel width of 3 mm, flow rate of 0 m^3/min, and heat input of 5 W.

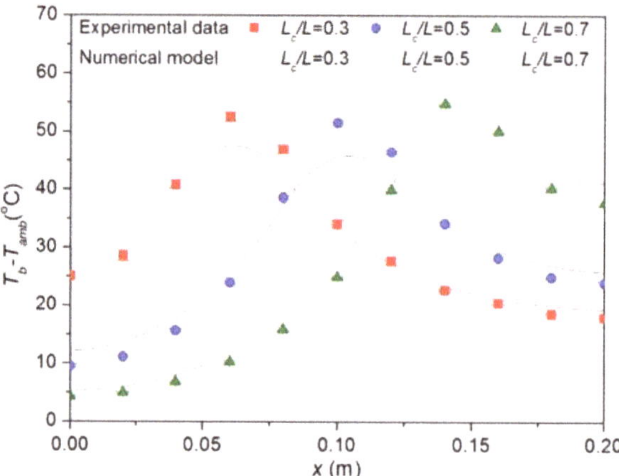

Figure A2. Comparison of temperature differences from experimental data and results of the numerical model for L_h/L of 0.2, channel width of 7 mm, and flow rate of 0.2 m^3/min.

References

1. Mudawar, I. Assessment of high-heat-flux thermal management schemes. *IEEE Trans. Compon. Packag. Technol.* **2001**, *24*, 22–141. [CrossRef]
2. Anandan, S.S.; Ramalingam, V. Thermal management of electronics: A review of literature. *Therm. Sci.* **2008**, *12*, 5–26. [CrossRef]
3. Yang, J.; Ning, S. Experimental and Numerical Study of Double-Pipe Evaporators Designed for CO_2 Transcritical Systems. *Processes* **2019**, *7*, 547. [CrossRef]
4. Zhou, H.; Chen, M.; Han, X.; Cao, P.; Yao, F.; Wu, L. Enhancement Study of Ice Storage Performance in Circular Tank with Finned Tube. *Processes* **2019**, *7*, 266. [CrossRef]
5. Xue, R.; Ruan, Y.; Liu, X.; Chen, L.; Liu, L.; Hou, Y. Numerical Study of the Effects of Injection Fluctuations on Liquid Nitrogen Spray Cooling. *Processes* **2019**, *7*, 564. [CrossRef]
6. He, M.; Li, Z.; Han, X.; Cabassud, M.; Dahhou, B. Development of a Numerical Model for a Compact Intensified Heat-Exchanger/Reactor. *Processes* **2019**, *7*, 454. [CrossRef]
7. Teertstra, P.; Yovanovich, M.M.; Culham, J.R. Analytical forced convection modeling of plate fin heat sinks. *J. Electron. Manuf.* **2000**, *10*, 253–261. [CrossRef]
8. Culham, J.R.; Muzychka, Y.S. Optimization of plate fin heat sinks using entropy generation minimization. *IEEE Trans. Compon. Packag. Technol.* **2001**, *24*, 159–165. [CrossRef]
9. Jonsson, H.; Moshfegh, B. Modeling of the thermal and hydraulic performance of plate fin, strip fin, and pin fin heat sinks-Influence of flow bypass. *IEEE Trans. Compon. Packag. Technol.* **2001**, *24*, 142–149. [CrossRef]
10. Iyengar, M.; Bar-Cohen, A. Least-energy optimization of forced convection plate-fin heat sinks. *IEEE Trans. Compon. Packag. Technol.* **2003**, *26*, 62–70. [CrossRef]
11. Kim, D.K.; Jung, J.; Kim, S.J. Thermal optimization of plate-fin heat sinks with variable fin thickness. *Int. J. Heat Mass Transf.* **2010**, *53*, 5988–5995. [CrossRef]
12. Muzychka, Y.S.; Yovanovich, M.M.; Culham, J.R. Influence of geometry and edge cooling on thermal spreading resistance. *J. Thermophys Heat Transf.* **2006**, *20*, 247–255. [CrossRef]
13. Musallam, M.; Johnson, C.M. Real-time compact thermal models for health management of power electronics. *IEEE Trans. Power Electron.* **2010**, *25*, 1416–1425. [CrossRef]
14. Yoon, Y.; Park, S.J.; Kim, D.R.; Lee, K.S. Thermal performance improvement based on the partial heating position of a heat sink. *Int. J. Heat Mass Transf.* **2018**, *124*, 752–760. [CrossRef]
15. Culham, J.R.; Khan, W.A.; Yovanovich, M.M.; Muzychka, Y.S. The influence of material properties and spreading resistance in the thermal design of plate fin heat sinks. *J. Electron. Packag.* **2007**, *129*, 76–81. [CrossRef]
16. Ellison, G.N. Maximum thermal spreading resistance for rectangular sources and plates with nonunity aspect ratios. *IEEE Trans. Compon. Packag. Technol.* **2003**, *26*, 439–454. [CrossRef]
17. Cho, E.S.; Choi, J.W.; Yoon, J.S.; Kim, M.S. Experimental study on microchannel heat sinks considering mass flow distribution with non-uniform heat flux conditions. *Int. J. Heat Mass Transf.* **2010**, *53*, 2159–2168. [CrossRef]
18. Yoon, S.H.; Saneie, N.; Kim, Y.J. Two-phase flow maldistribution in minichannel heat-sinks under non-uniform heating. *Int. J. Heat Mass Transf.* **2014**, *78*, 527–537. [CrossRef]
19. Lelea, D. The heat transfer and fluid flow of a partially heated microchannel heat sink. *Int. Commun. Heat Mass Transf.* **2009**, *36*, 794–798. [CrossRef]
20. Toh, K.C.; Chen, X.Y.; Chai, J.C. Numerical computation of fluid flow and heat transfer in microchannels. *Int. J. Heat Mass Transf.* **2002**, *45*, 5133–5141. [CrossRef]
21. Emekwuru, N.G.; Hall, F.R.; Spence, P.J. Partially heated heat sinks in a zero-bypass. *Int. Commun. Heat Mass Transf.* **2012**, *39*, 343–349. [CrossRef]
22. Lee, M.; Kim, H.J.; Kim, D.K. Nusselt number correlation for natural convection from vertical cylinders with triangular fins. *Appl. Therm. Eng.* **2016**, *93*, 1238–1247. [CrossRef]
23. Kim, D.K.; Han, I.Y.; Kim, S.J. Study on the steady-state characteristics of the sensor tube of a thermal mass flow meter. *Int. J. Heat Mass Transf.* **2007**, *50*, 1206–1211. [CrossRef]

24. Liu, D.; Garimella, S.V. Analysis and optimization of the thermal performance of microchannel heat sinks. *Int. J. Numer. Methods Heat Fluid Flow* **2005**, *15*, 7–26. [CrossRef]
25. Incropera, F.P.; Lavine, A.S.; Bergman, T.L.; DeWitt, D.P. *Fundamentals of Heat and Mass Transfer*, 6th ed.; John Wiley & Sons: Hoboken, NJ, USA, 2007; pp. 153–155.

Article

Heat Flux Estimation at Pool Boiling Processes with Computational Intelligence Methods

Erdem Alic [1,*] , **Mehmet Das** [2] **and Onder Kaska** [3]

[1] Mechanical Engineering Department, University of Kahramanmaras Sutcu Imam, 46400 Kahramanmaras, Turkey

[2] Vocation High School of Ilic Dursun Yildirim, Erzincan Binali Yildirim University, Ilic, 24700 Erzincan, Turkey; mdas@erzincan.edu.tr

[3] Mechanical Engineering Department, University of Osmaniye Korkut Ata, 80000 Osmaniye, Turkey; onderkaska@osmaniye.edu.tr

* Correspondence: ealic@ksu.edu.tr

Received: 27 March 2019; Accepted: 6 May 2019; Published: 17 May 2019

Abstract: It is difficult to manually process and analyze large amounts of data. Therefore, to solve a given problem, it is easier to reach the solution by studying the data obtained from the environment of the problem with computational intelligence methods. In this study, pool boiling heat flux was estimated in the isolated bubble regime using two optimization methods (genetic and artificial bee colony algorithm) and three machine learning algorithms (decision tree, artificial neural network, and support vector machine). Six boiling mechanisms containing eighteen different parameters in the genetic and the artificial bee colony (ABC) algorithms were used to calculate overall heat flux of the isolated bubble regime. Support vector machine regression (SVMReg), alternating model tree (ADTree), and multilayer perceptron (MLP) regression only used the heat transfer equation input parameters without heat transfer equations for prediction of pool boiling heat transfer over a horizontal tube. The performance of computational intelligence methods were determined according to the results of error analysis. Mean absolute error (MAE), root mean square error (RMSE), and mean absolute percentage error (MAPE) error were used to calculate the validity of the predictive model in genetic algorithm, ABC algorithm, SVMReg, MLP regression, and alternating model tree. According to the MAPE error analysis, the accuracy values of MLP regression (0.23) and alternating model tree (0.22) methods were the same. The SVMReg method used for pool boiling heat flux estimation performed better than the other methods, with 0.17 validation error rate of MAPE.

Keywords: boiling; computational intelligence techniques; heat flux; optimization

1. Introduction

Pool boiling processes are important heat transfer mechanisms in many engineering applications [1], especially in chemistry, mechanical engineering processes, refrigeration, gas separation, etc. [2]. The formation and removal of vapor bubbles from the solid–liquid interface can be explained by boiling. In the literature, boiling heat transfer studies can be divided into two groups: (1) flow boiling; and (2) pool boiling [3,4]. Boiling allows the transfer of large amounts of heat energy at low-temperature differences. The boiling event has a wide range of applications. The major areas of application include nuclear power plants, rocket motors, refrigeration industry, boilers, steam power units, process industry, and evaporators. Although many investigations are reported on the boiling mechanism, the physical mechanism of boiling has not yet been fully elucidated, even in the case of running water [5]. Many investigators have improved correlation for calculating boiling heat flux [6]. These correlations are calculated for nucleate pool boiling heat flux to nearly 50% error [7–10]. Nowadays, many investigators study optimization and ANN for heat transfer prediction [11–14]. Das and Kishor studied the heat transfer coefficient in pool boiling of distilled water. They compared the results of the zero-order adaptive fuzzy model and adaptive neuro-fuzzy inference system (ANFIS function) [15]. Swain and Das used the computational intelligence methods for the flow boiling heat transfer coefficient [16]. Barroso-Maldonado et al. studied cryogenic forced boiling. They compared ANN to three conventional correlations [17].To calculate heat transfer in fluids, some researchers have developed models using computational fluid dynamics [18,19].

In recent years, many researchers have studied the optimization of the heat transfer of the pool boiling [1,20,21]. Many researchers have predicted heat flux with computational intelligence methods. Table 1 depicts the conditions under which the boiling heat transfer is calculated, the algorithms used in heat flux estimation, which error measures are used to determine the accuracy of the predictive models and the error analysis results. There are generally two types of computational intelligence methods: (1) white-box techniques; and (2) black-box techniques. Optimization techniques, such as genetic and ABC, are white-box techniques, while artificial intelligence techniques, such as ANN, DT, and MLP, are black-box techniques.

In this study, the pool boiling heat flux was calculated by optimizing semi-empirical correlations. Then, heat flux estimation was realized using computational intelligence methods considering the parameters used in the calculation of conventional correlations. These methods were also compared with well-known correlations. To the authors' best knowledge, this study contributes to the heat flux estimation for pool boiling literature by using black-box techniques for the first time.

2. Pool Boiling Mechanisms in Isolated Bubble Regime Region

The boiling process occurs when the temperature of the solid surface to which the liquid contacts exceeds the saturation temperature corresponding to the pressure of the liquid. Boiling process is described visually in Figure 1. In Figure 1, the pipe diameter and length are 21 mm and 105 mm, respectively. The experimental heat flow is about 10–80 kW/m^2. Four different materials, copper, brass, aluminum, and steel, are used as heater surface. The surface roughness is 30–360 mm and the conditions are atmospheric pressure. The vessel of the boiling volume is 0.003 m^3. Water and ethanol were used for a boiling liquid. As seen in Figure 1, the first bubble was boiled in the boiling core in the isolated bubble regime. However, the bubble did not reach the free surface. In an isolated bubble regime, the boiling core was analyzed for boiling the pool on a horizontal tube heater [20].

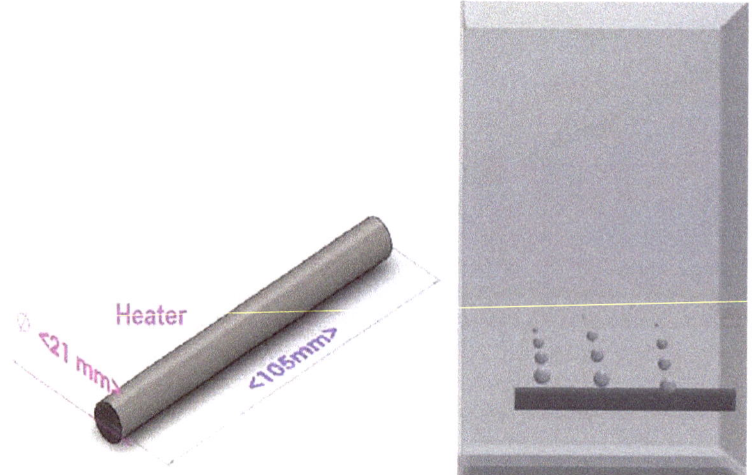

Figure 1. The schematic problem description.

In the region of the isolated vapor bubble, the vapor bubbles do not emerge on the fluid surface and become condensed again by condensation. Although the temperature difference between the fluid and the surface is low in this region, the amount of heat transferred is very high. According to the Fazel study [20], the isolated bubble regime had six mechanisms (microlayer evaporating, transient conduction, bubble super-heating, sliding bubbles for transient conduction, radial forced convection, and natural convection). The model used for optimization in the study is a semi-empirical model that estimates heat flux by a Genetic Algorithm [20]. Fazel's model, although improved at the isolated bubble regime region, has the following limitations: (1) the heat transfer is one dimensional; (2) bubbles are adopted in a spherical shape; (3) tThe heater temperature is constant; (4) bubbles are isolated, i.e., there are no bubble interactions; and (5) the physical properties do not change [20]. The boiling heat transfer model as the sum of the six mechanisms is written below.

Microlayer evaporating equation [22]:

$$h^{mic} = \frac{\pi}{6} d^3 \rho_v h_{fg} \frac{N}{A} f \tag{1}$$

Transient conduction equation [22]:

$$(h)^{trns} = 2\sqrt{\frac{\rho_l k_l c_{pl}}{3 t_w}} \Delta T \frac{N}{A} P_1 \frac{\pi}{4} d^2 P_0 \tag{2}$$

$$P_0 = \frac{t_w}{\tau} = t_w f \tag{3}$$

Bubble super-heating equation [20]:

$$(h)^{bb} = 2\sqrt{\frac{\rho_v k_v c_{pv}}{3 t_g}} \Delta T \frac{N}{A} P_2 \frac{\pi}{4} d^2 (1 - P_0) \tag{4}$$

Sliding bubbles for transient conduction equation [20]:

$$l_s d = P_0 \left(\frac{N}{A}\right)^{-\frac{1}{2}} d \tag{5}$$

$$(h)^{stc} = 2\sqrt{\frac{\rho_l k_l c_{pl}}{3 t_w}} \Delta T \frac{N}{A} P_3 d \left(\frac{N}{A}\right)^{-\frac{1}{2}} P_0 \tag{6}$$

Radial forced convection equation [20]:

$$u^{radial} = \frac{d/2}{\tau_w} \tag{7}$$

$$Nu_r = 0.453 Re_r^{\frac{1}{2}} . Pr^{\frac{1}{3}} \tag{8}$$

$$Re_r = \frac{\rho_l u^{radial}(d/2)}{\mu_l} \tag{9}$$

$$\overline{Nu_r} = \frac{\int_0^r Nu_r 2\pi r dr}{\int_0^r 2\pi r dr} = (4/3) Nu_r \tag{10}$$

$$A_{a_{rfc}} = P_4 \frac{N}{A} \frac{\pi}{4} d^2 (1 - P_0) \tag{11}$$

$$h^{radf} = h_r P_4 \frac{N}{A} \frac{\pi}{4} d^2 (1 - P_0)(T_w - T_b) \tag{12}$$

Natural convection equation [23,24]:

$$h^{Nc} = \alpha_{Nc} c_{Nc} (T_w - T_b) \tag{13}$$

$$\overline{Nu}_{OD} = \left(0.6 + \frac{0.387 \cdot (Gr_{OD} \cdot Pr)^{1/6}}{\left(1 + \left((0.559/Pr)^{9/16}\right)\right)^{8/27}} \right)^2 \tag{14}$$

$$c_{nc} = 1 - c_1 - c_2 - c_4 \tag{15}$$

$$c_1 = \frac{N}{A} \frac{\pi}{4} d^2 \left[P_0 \frac{P_1 + P_2 + P_3 + P_4}{4} + (1 - P_0) \right] \tag{16}$$

$$c_2 = \frac{\pi}{4} d^2 P_5 \frac{N}{A} \tag{17}$$

$$c_3 = \sqrt{\frac{N}{A}} P_3 d \tag{18}$$

$$c_4 = P_0(c_3 - c_2) \tag{19}$$

The above equations and experimental data were taken from Fazel's work (see the article for more details) [20]. The pool boiling heat transfer is affected by many parameters that are easily obtained from correlations given in the literature: the wall temperature (T_w), evaporation temperature (T_b), bubble departure diameter (d), bubble frequency (f), nucleation site density (NA), latent heat vaporization (h_{fg}), liquid density (ϱ_l), vapor density (ϱ_v), liquid heat capacity (Cp_l), vapor heat capacity (Cp_v), dynamic viscosity μ , Prandtl number (Pr), liquid thermal conductivity (k_l), Grashoff number (Gr), and vapor thermal conductivity (k_v). These parameters are used to calculate the pool boiling heat flux (q/A). q/A is the total of the six mechanisms of heat flux equations. These parameters are available for computational intelligence technique as well.

Table 1. The predictive models of heat flux in the literature.

Conditions	Data	Method	Error Analysis	Result	Reference
Round tube uniform heat oscillation	513	ANN	MSE	0.2	Kim et al. [25]
Down-stream conditions, Vertical round tube	513	ANN	MSE	0.25	Kim et al. [26]
PWR steady-state	60	SVM	R²	0.65	He and Lee [27]
Concentric-tube open thermosiphon	381	SVM	RMSE	0.067	Cai [28]
Bubble column reactors	366	SVR	AARE	7.05%	Gandhi and Joshi [29]
Steam-water flows in pipes	3000	ANN	MSE	0.75	Nafey [30]
Wall insulation surface	342	ANN	RMSE	0.0631	Najafi and Woodbury [31]
Vertical smooth tube	368	MLP	R²	0.992	Balcilar et al. [32]
Mini-Channel	319	ANN	R²	0.998	Parveen et al. [33]

3. Computational Intelligence Methodology

In a mathematical approach, most optimization methods investigate the places where the function is zero and the places where the derivative is zero. The derivative calculation is not always an easy task. Many of the technical problems can be formulated to find their roots. However, some optimization methods fail to find these roots. Another challenge in optimization is determining whether a result is a global or local solution. Such problems are solved either by a linear approach or by limiting the bounds of the optimization domain. In this study, five different computational intelligence methodologies were selected to estimate pool boiling heat transfer in isolated bubble regime.

3.1. Genetic Algorithm

Genetic Algorithm is the first and most known of the evolutionary calculation algorithms. To understand the terminology of the genetic algorithm (GA), it is necessary to understand natural selection. When observing the world, natural selection comes to the fore in events. The enormous organisms and complexity of these organisms are the subject of investigation and research. It can be questioned why organisms are like this and how they come to this stage. The level of adaptation and suitability has become a sign of long-term survival in the world. The process of evolution is a great algorithm that allows the most appropriate life conditions. If an organism has the intelligence and ability to change the environment, the global maximum can be achieved in life [34]. This algorithm externalizes the process of natural selection in which pertinent individuals are selected for reproduction to produce progeny of the next generation. General flow chart of the genetic algorithm is given in Figure 2.

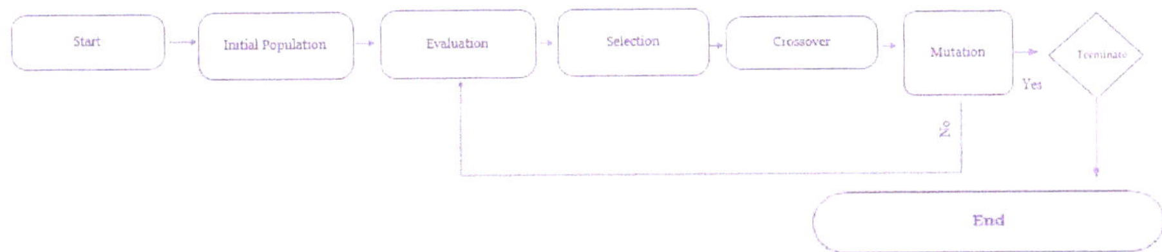

Figure 2. Genetic algorithm steps.

3.2. ABC Algorithm

ABC algorithm is the modeling of bees food search mechanisms. The bees live in colonies. The bees colony includes three groups of bees: onlookers, scouts, and employed bees. Some of the colony consists of employed artificial bees and the others contain onlookers. For each food source, there is solely an employed bee [35,36]. Figure 3 shows the main steps for the ABC algorithm. In this study, the ABC algorithm and the GA algorithm were used with the same mathematical model and bounds; however, their configurations were different. Both algorithm configurations are shown in Table 2.

Table 2. Configuration of both algorithms.

	Genetic	ABC
Number of variable	6	6
Bounds	[0 0.1 0.1 0.1 0.1 0.1] & [1 5 5 5 3 4]	[0 0.1 0.1 0.1 0.1 0.1] & [15 5 5 3 4]
Population type	double vector	
Population size	150	150
Creation function	uniform	
Fitness scaling	top	
Mutation function	adaptive feasible	
Crossover	constraint dependent	
Migration	forward	
Hybrid function	none	
Stopping criteria	default	
Generations	50	50
Stall test	average change	
Others	default	
Use random states from previous run	sign	
User function evaluation	in serial	
Run time	30	30

The optimization model was based on 18 parameters: (1) q/A; (2) T_w; (3) T_b; (4) d; (5) NA; (6) f; (7) ϱ_l; (8) ϱ_v; (9) h_{fg}; (10) Cp_l; (11) k_l; (12) k_v; (13) Cp_v; (14) μ_l; (15) μ_v; (16) Gr; (17) Prandtl; and (18) Ra roughness. Some of these parameters were obtained from experiments: (1) T_w; (2) T_b; (3) d; (4) NA; (5) Ra; and (6) f. The rest of the datasets were obtained from an EES package program: (1) ϱ_l; (2) ϱ_v; (3) h_{fg}; (4) Cp_l; (5) k_l; (6) k_v; (7) Cp_v; (8) μ_l; (9) μ_v; and (10) Prandtl. If these datasets were used as the input parameter, for the computational intelligence algorithms, boiling heat flux (q/A) could be estimated. The boiling fluid thermophysical properties were evaluated at the arithmetic mean of the saturated fluid and heater surface temperature, T_f defined by Equation (20).

$$T_f = \frac{T_w + T_b}{2} \tag{20}$$

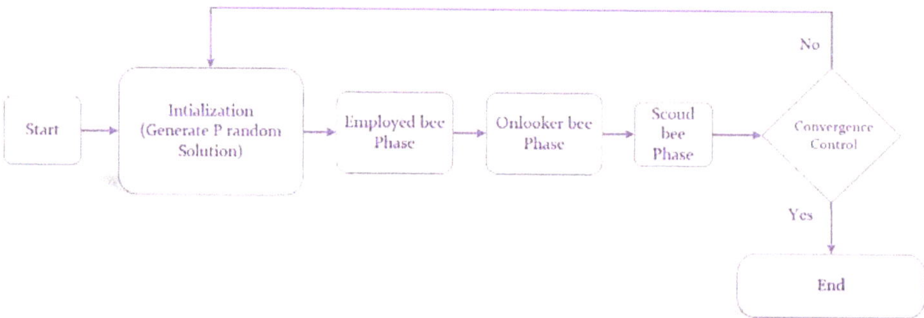

Figure 3. ABC algorithm steps.

3.3. Support Vector Machine Regression

Support vector networks, which are a variety of universal feeder networks, were developed by Vapnik and Cortes [37] to classify data and are generally referred to as support vector machines (SVM) in the literature. The SVM-based model for regression is called the support vector regression (SVMReg) [38]. SVR uses not only empirical risk minimization but also the principle of structural risk minimization, which is intended to reduce the upper limit of the generalization error, compared to traditionally controlled learning methods of neural networks. Thanks to this principle, the SVR has good generalization performance for previously untested test data using the learned input–output relationship during the training phase. Consider the expression vector $x_s \in R^n$ for the problem of approach to a continuous-valued function. The expression $D = \{(x_s, y_s) \,|s \in \{1, 2, ..., N\}\,\}$, which is a set of N numbers, indicating the $y_s \in R$ output (target) value. The aim of the regression analysis is to determine a mathematical function to accurately predict the desired (target) outputs ($y_s \in R$). The regression problem can be classified as linear and nonlinear regression problems. Since the problem of nonlinear regression is more difficult to solve, SVMReg is mainly developed for the solution of nonlinear regression problem [39]. To solve the nonlinear regression problem, the SVM carries the training data in the "i" input space $\varphi(x)R^n \to R^m (m > n)$ to the higher-dimensional space $\{(\varphi(x_s), y_s) \,|s \in \{1, 2, ..., N\}\,\}$ with the help of a nonlinear function and applies linear regression in this space. In this case, the mathematical representation of the linear function obtained to find the best regression is as follows [39]:

$$f(x, w) = \sum_{s=q}^{N} w_s \varphi(x_s) + b = w^T \varphi(x) + b \tag{21}$$

where $w \in R^m$ represents the model parameter vector and $b \in R$ represents the deviation term in the vertical axis. Thus, the linear regression obtained by the inner product between ω and $\varphi(x)$ in the higher-dimensional space corresponds to the nonlinear regression in the input space (Figure 4). The objective function of the SVR, which performs linear regression in high-dimensional space, is usually composed of a ϵ-insensitive loss function and minimization of the parameters representing the model in Equation (22).

$$\min_{w \in R^w, b \in R} J(w, b) = \frac{1}{2} \sum_{s=1}^{N} \|w\|^2 + C \sum_{s=1}^{N} L_\varepsilon(y_s, f(x_s)) \tag{22}$$

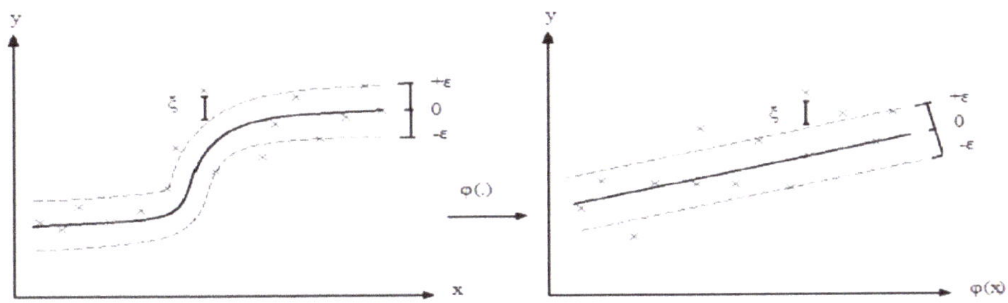

Figure 4. Using the nonlinear function $\varphi(.)$ mapping training examples in the input space to a high dimension where they are linear.

Here, the first term $\|w\|^2/2$ represents the square of the Euclidean norm of the model parameters, the second term $L\varepsilon(y_s, f(x_s))$ is the experimental error (loss) function, and the $C \in R^+$ is a positive constant number. The task of C is to maintain a balance between the experimental error and the extreme compatibility of the model with the training data. Small C values give more importance to the optimization problem in contrast to the experimental error, while the higher C values give more importance to the reduction of experimental education error than the norm of ω [39]. SVM regression computational intelligence method was used to create a predictive model of q/A values calculated by experimental data. This method was done with the SMOreg toolbox in WEKA 3.8.3. WEKA is open source software. There are many algorithms in this software, which include classification, estimation and clustering rules. It is necessary to define the kernel function to be used for a classification to be performed by SVM and optimum parameters of this function [40]. The most widespread used radial basis function (RBF) kernels in the literature are presented together with formulas and parameters in Equation (23). Batch size is 1000. "C" is 200.0. Filter type is standardized.

$$K(x.y) = e^{-\gamma \|(x-xi)\|^2} \tag{23}$$

3.4. Multilayer Perceptron

Artificial intelligence has been brought to science through long-term studies to model the human brain. Then, the artificial neural networks (ANN) method was developed by means of these studies. The ANN technique achieved reliable results in nonlinear equation solutions and its use has become increasingly widespread over time. In the ANN method, models with a multi-layer perceptron (MLP) are generally used for classification and regression approaches. The software application of the MLP neural networks is used in the algorithm development phase and in cases where parallel-low delay approaches are not required. In many applications in the literature, rapid data processing, and low delays are required by the ANN method. To supply this need, MLP is used as an ANN method consisting of multiple neural layers in a feed-through network. The MLP consists of three or more layers consisting of one inlet, one outlet, and one or more hidden layers. Because the MLP is a fully connected network, each neuron contained in each layer is associated with the next layer with a certain weight value. The MLP method uses a controlled learning method called backpropagation.

In MLP, the weight function is defined in the training phase of the neural network [41]. As a middle layer, six hidden layers were created and the best solutions were tried by changing the number of intermediate layers. The structure of the generated MLP model is given in Figure 5.

Prediction of q/A values with MLP was done using WEKA 3.8.3 software. The MLP network structure used for estimation of q/A is shown in Table 3.

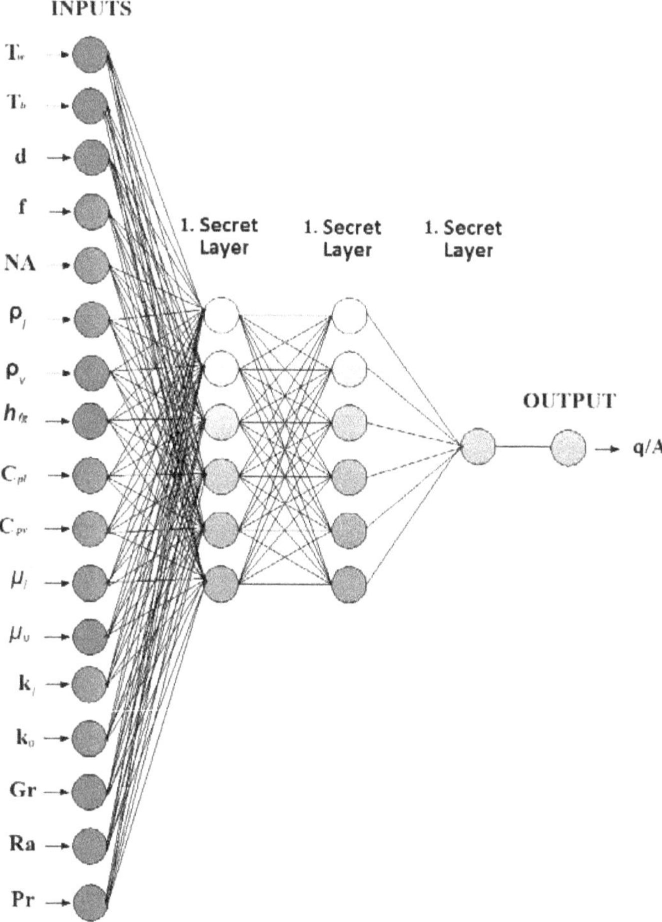

Figure 5. MLP structure.

Table 3. The Network structure of MLP.

Network Structure of MLP Regression	
Number of Secret Layers	3
Number of Neurons in Layers	6-6-1
Weight Ratings	Random
Activation Function	Softplus
Transfer Function	Tangent Sigmoid Transfer
Learning Function	Backpropagation

3.5. Alternating Decision Tree

Decision trees, which are a strong regression method, have a clear concept description for a dataset. The decision tree learning method is a popular method because of its fast data processing capability and because it produces successful performance predictive models [42,43]. The alternative decision tree (ADTree) method consists of decision nodes and prediction nodes. Each action in the decision nodes indicates the result. The prediction nodes contain a single number value. The ADTree method always contains prediction nodes, which consist of both root and leaves. When regression or classification is done by the ADTree method, the paths of all decision nodes and prediction nodes are monitored [44]. In the ADTree method, the learning algorithm must be $1 <= i <= n$. In this expression, n is provided via a sample $(x_i; y_i).x_i$. In this sample, xi indicates an attribute value indicating the vector and yi indicates the target value. For this dataset, when a different vector x is entered, this model is used to estimate the value corresponding to the y value. The purpose of the ADTree model is to minimize the error between the actual value and the estimated value. The ADTree method uses the basic algorithm of incremental regression by using the advanced stepwise additive model at the stage of learning additive model trees [45]. If a model consisting of k base model is created,

$$F_k(\vec{x}) = \sum_{j=1}^{k} f_j(\vec{x}) \tag{24}$$

the error squared on a progressive state,

$$\sum_{i=1}^{n} \left(F_k(\vec{x}_i) - y_i \right)^2 \tag{25}$$

is minimized through n training samples.

As input data for the all regression analysis model, the following were used: (1) T_w; (2) T_b; (3) d; (4) NA; (5) f; (6) ϱ_l; (7) ϱ_v; (8) h_{fg}; (9) Cp_l; (10) k_l; (11) k_v; (12) Cp_v; (13) μ_l; (14) μ_v; (15) Gr; (16) Prandtl; and (17) Ra roughness values. Boiling heat flux (q/A) was used as output data.

All numerical method errors were calculated for MAE, MAPE, and RMSE. All error calculation methods are shown in Table 4.

Table 4. Some error equations.

Accuracy Criteria	Formulas	Parameters
		P: Predicted Value
MAE	$\dfrac{\lvert P1-A1\rvert + \dots + \lvert Pn-An\rvert}{n}$	A: Actual Value
		n: Total Estimated Value P: Predicted Value
RMSE	$\sqrt{\dfrac{(P1-A1)^2 + \dots + (Pn-An)^2}{n}}$	A: Actual Value
		n: Total Estimated Value d: Predicted Value
MAPE	$\sum_{p}^{P} \left\lvert \dfrac{d_p - z_p}{d_p} \right\rvert \times \dfrac{100}{P}$	z: Actual Value
		P: Total Estimated Value

The test dataset is used to determine the generalization capability of the generated tree for a new dataset. A test data coming from the root of the tree enters the tree structure created with the training dataset. This new data tested in the root is sent to a lower node according to the test result. This process is continued until it reaches a specific leaf of the tree. There is only one way or a single decision rule from root to every leaf [46]. The working principle of the decision tree method, which is a computational intelligence method, is simply shown in Figure 6. Figure 6 shows a simple tree structure consisting of four-dimensional attribute values of three classes. In the Figure 6, the xi parameter shows the values of the attribute. The parameters a, b, c, and d show the threshold values in tree branches. Parameters A, B, and C show the class label values [47].

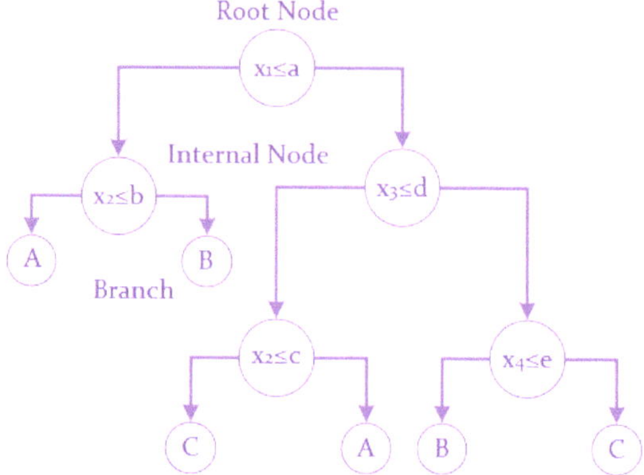

Figure 6. A decision tree structure consisting of three classes with four-dimensional property space.

4. Results

Five different computational intelligence methods were applied to predict pool boiling heat transfer phenomena. Genetic and ABC algorithms are both white-box algorithms, in which the internal structure, design, and implementation of the item being tested are known to the tester. All parameters required for algorithms were obtained from Fazel [20]. The black-box technique is the opposite of the white-box technique and its algorithms cannot be changed. They can only be partially modified for the prediction, for example the learning rate, momentum, batch size, and Figure 7 shows the predicted output of the SVMReg. In the figure, it can be seen that the predicted output of the SVMReg and experimental data are mostly in agreement.

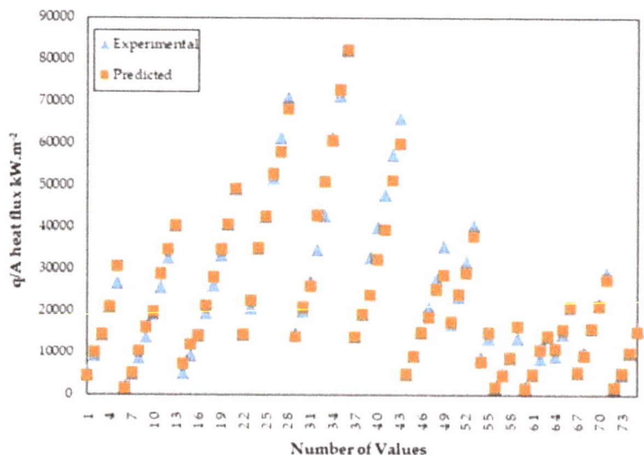

Figure 7. The SVMReg forecast.

Figure 8 shows a comparison of the black-box techniques to white-box techniques. The apsis is the experimental results, and the ordinates are the predicted models. The figure shows that both algorithms predictions were nearby the 0.25 mean absolute percentage error of the actual result. MLP and Alternating tree had unequal distributions, whereas SVMReg prediction was slightly more stable than the others. Therefore, its mean absolute percentage error was less than all other methods. SVMReg model performed better than the other methods.

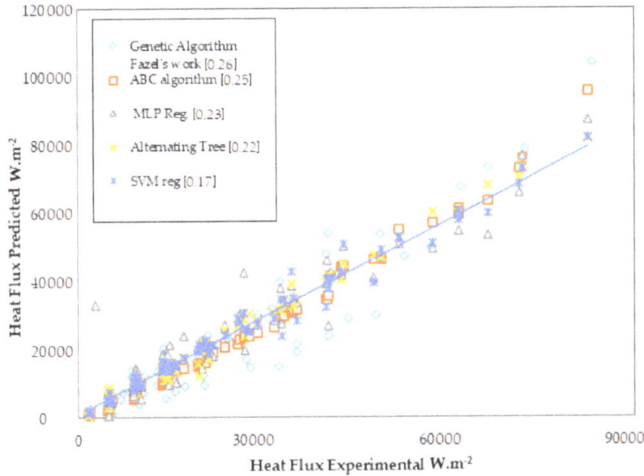

Figure 8. The comparison of black-box techniques to white-box techniques.

Seventeen attributes (1275) were used for analysis. An attribute q/A (75) was selected to be used as the solution class. In this study, the ten-fold cross-validation technique was applied to process the data with less error rate in machine learning algorithms. Cross-validation is a technique used in model selection to better estimate the error of a test in a machine learning model. In cross-validation, the training data are divided into subsets. A subset is used for training, and the remaining sets are used for validation. This process is repeated for all subsets in a crossway. This is done for k presets. In the literature, ten-cross validation can be seen in many articles [33,48]. Data are divided into k pieces of equal length and evaluated k times. The mean absolute error (MAE), root mean squared error (RMSE) and mean percentage error (MAPE) are shown in Table 5.

Table 5. Error rates for all methods.

Model Name	MAE	RMSE	MAPE
Genetic Algorithm [20].	5.1924	6.4435	0.26
ABC algorithm [21].	4.9519	5.23758	0.25
MLP.	3.5131	4.6221	0.23
SVMReg	2.2716	4.0026	0.17
Alternating Tree Model	3.3826	4.2818	0.22

Instant heat flux has to be estimated in pool boiling. This requirement is mainly for determining the boiling heat transfer coefficient. Figure 9 depicts the results of the well-known correlations and computational intelligence techniques used in the study. It is clearly seen that the computational intelligence techniques performed better than the correlations developed. Especially, SVMReg predicted q/A error rate with a MAPE nearby 0.17. This error rate was the minimum value presented in this study. In this analysis, some parameters were obtained as a result of experiments. The other parameters were obtained from EES. However, they could make better predictions than the many correlations used today for the development of data. In the near future, computational intelligence techniques can be used more in predicting the heat transfer of boiling phenomena. Some researchers supported this view in their works [49–51].

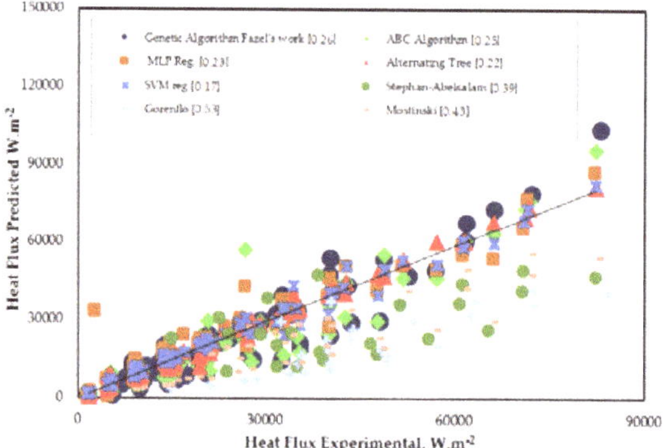

Figure 9. Comparison of some well-known correlation and computational intelligence techniques.

In this study, heat flux estimation was performed using computational intelligence methods during boiling in the isolated bubble regime region. After this study, these methods of computational intelligence can be tried in the boiling zone, where the steam bubble flows to the surface of the fluid, for the transition to boiling, and in the film boiling. In addition, these methods can be tried to estimate the heat transfer coefficient in boiling heat transfer. It is thought that these methods will be successful in estimating the heat transfer coefficient. Thus companies that produce heat exchangers for the industry could benefit from these algorithms in their heat exchanger capacity estimation. The prediction results obtained by these algorithms were found to be better than those obtained with the known correlations.

5. Conclusions

The novelty of this study was to predict the boiling heat flux of a pool by using black-box techniques. Pool boiling heat transfer was predicted with computational intelligence techniques. These computational intelligence methods were Genetic algorithms, ABC algorithms, SVM, DT, and MLP. The predicted heat flux was compared to some well-known correlations. The white-box techniques performance (Genetic and ABC) was limited to the used empirical model, whereas predictions made by black-box techniques (SVM, DT, and MLP) were more successful. Validation error (MAPE) rate of the models were: GA, 0.26; ABC, 0.25; MLP, 0.23; DT, 0.22; and SVM, 0.17. SVMReg was proposed as the best of the models used in the study to predict the heat transfer phenomenon in pool boiling. This study also showed the basics of how to use computational intelligence techniques in engineering calculation programs. By making an addendum to engineering equation solvers such as EES, the ability to use computational intelligence techniques can be improved, more data can be obtained with different boiling techniques, and less erroneous predictive models can be obtained using different computational intelligence methods.

Author Contributions: O.K. supervised all aspects of the research. E.A. and M.D. developed computational intelligence methods and wrote the paper.

Funding: There was no funding for this study.

Acknowledgments: There are no acknowledgments for this study.

Conflicts of Interest: The authors declare no conflict of interest.

Abbreviations

The following abbreviations are used in this manuscript:

A	Area (m^2)
$AARE$	Average Absolute Relative Error
$ADTree$	Alternative Decision Tree
ANN	Artificial neural network
C	Batch size
c_1	The particular area engaged by the bubbles over heater surface area
c_2	The particular area engaged by the sliding bubbles over heater surface area
c_3	The particular area over the area at which transient heat conduction takes
c_4	Various of c_3 and c_2
C_p	Heat capacity
DT	Decision tree
E	The relative error (%)
f	The frequency of bubble separation (Hz)
Gr	Dimensionless Grashoff number
h	Heat flux watt per square metre (W m^{-2})
h_{fg}	Specific heat of vaporization (J kg^{-1})
k	Thermal conductivity (W m^{-1} K^{-1})
MAE	Mean absolute error
$MAPE$	Mean absolute percentage error
MSE	Mean squared error
MLP	Multilayer perceptatron
NA	Nucleation site density
Nc	Natural convection
Nu	The Nusselt number
P	Optimization parameter
q/A	Sum of heat flux (W m^{-2})
RBF	Radial basis function
Re	Reynolds number
$RMSE$	Root mean square error
SVM	Support vector machine
SVR	Support vector regression
T	Temperature (K)
Greek symbols	
ϱ	Density (kg m^{-3})
μ	Viscosity (kg m^{-1} s^{-1})
ω	Pearson width parameters
γ	Kernel dimension
Subscripts or superscripts	
l_s	Sliding length (m)
b	Bulk
bb	Bubble super-heating
d	The diameter of bubble separation (m)
v	Vapor
l	Liquid
mic	Micro-layer evaporation
OD	Heater outside diameter (m)
$radf$	Radial forced convection
s	Saturation temperature
th	Thermocouple location
$trns$	Transient conduction
w	Wall

References

1. Ahn, H.S.; Sathyamurthi, V.; Banerjee, D. Pool Boiling Experiments ona Nano-structured surface. *IEEE Trans. Compon. Packag. Technol.* **2009**, *32*, 156–165.
2. Kandlikar, S.G. *Handbook of Phase Change: Boiling and Condensation*; CRC Press: Boca Raton, FL, USA; Routledge: Abingdon, UK, 1999.
3. Yan, B.H.; Wang, C.; Li, R. Corresponding principle of critical heat flux in flow boiling. *Int. J. Heat Mass Transf.* **2019**, *136*, 591–596. [CrossRef]
4. Colgan, N.; Bottini, J.L.; Martínez-Cuenca, R.; Brooks, C.S. Experimental study of critical heat flux in flow boiling under subatmospheric pressure in a vertical square channel. *Int. J. Heat Mass Transf.* **2019**, *130*, 514–522. [CrossRef]
5. Chai, L.H.; Peng, X.F.; Wang, B.X. Nonlinear aspects of boiling systems and a new method for predicting the pool nucleate boiling heat transfer. *Int. J. Heat Mass Transf.* **2000**, *43*, 75–84. [CrossRef]
6. Zahedipoor, A.; Faramarzi, M.; Eslami, S.; Malekzadeh, A. Pool boiling heat transfer coefficient of pure liquids using dimensional analysis. *J. Part. Sci. Technol.* **2017**, *3*, 63–69.
7. Gorenflo, D. *Pool Boiling, VDI-Heat Atlas*; VDI-Verlag: Dusseldorf, Germany, 1993.
8. Stephan, K.; Abdelsalam, M. Heat-transfer correlations for natural convection boiling. *Int. J. Heat Mass Transf.* **1980**, *23*, 73–87. [CrossRef]
9. McNelly, M.J. A correlation of rates of heat transfer to nucleate boiling of liquids. *J. Imp. Coll. Chem. Eng. Soc.* **1953**, *7*, 18–34.
10. Mostinski, I.L. Application of the rule of corresponding states for calculation of heat transfer and critical heat flux. *Teploenergetika* **1963**, *4*, 66–71.
11. Chang, W.; Chu, X.; Fareed, A.F.B.S.; Pandey, S.; Luo, J.; Weigand, B.; Laurien, E. Heat transfer prediction of supercritical water with artificial neural networks. *Appl. Therm. Eng.* **2018**, *131*, 815–824. [CrossRef]
12. Esfe, M.H.; Arani, A.A.A.; Badi, R.S.; Rejvani, M. ANN modeling, cost performance and sensitivity analyzing of thermal conductivity of DWCNT-SiO2/EG hybrid nanofluid for higher heat transfer: An experimental study. *J. Therm. Anal. Calorim.* **2018**, *131*, 2381–2393. [CrossRef]
13. Hojjat, M. Modeling heat transfer of non-Newtonian nanofluids using hybrid ANN-Metaheuristic optimization algorithm. *J. Part. Sci. Technol.* **2018**, *3*, 233–241.
14. Harsha Kumar, M.K.; Vishweshwara, P.S.; Gnanasekaran, N.; Balaji, C. A combined ANN-GA and experimental based technique for the estimation of the unknown heat flux for a conjugate heat transfer problem. *Heat Mass Transf.* **2018**, *54*, 3185–3197.
15. Das, M.K.; Kishor, N. Adaptive fuzzy model identification to predict the heat transfer coefficient in pool boiling of distilled water. *Expert Syst. Appl.* **2009**, *36*, 1142–1154. [CrossRef]
16. Swain, A.; Das, M.K. Artificial intelligence approach for the prediction of heat transfer coefficient in boiling over tube bundles. *Proc. Inst. Mech. Eng. Part C* **2014**, *228*, 1680–1688. [CrossRef]
17. Barroso-Maldonado, J.M.; Montañez-Barrera, J.A.; Belman-Flores, J.M.; Aceves, S.M. ANN-based correlation for frictional pressure drop of non-azeotropic mixtures during cryogenic forced boiling. *Appl. Therm. Eng.* **2019**, *149*, 492–501. [CrossRef]
18. Moradikazerouni, A.; Afrand, M.; Alsarraf, J.; Wongwises, S.; Asadi, A.; Nguyen, T.K. Investigation of a computer CPU heat sink under laminar forced convection using a structural stability method. *Int. J. Heat Mass Transf.* **2019**, *134*, 1218–1226. [CrossRef]
19. Moradikazerouni, A.; Afrand, M.; Alsarraf, J.; Mahian, O.; Wongwises, S.; Tran, M.-D. Comparison of the effect of five different entrance channel shapes of a micro-channel heat sink in forced convection with application to cooling a supercomputer circuit board. *Appl. Therm. Eng.* **2019**, *150*, 1078–1089. [CrossRef]
20. Alavi Fazel, S.A. A genetic algorithm-based optimization model for pool boiling heat transfer on horizontal rod heaters at isolated bubble regime. *Heat Mass Transf./Waerme- und Stoffuebertragung* **2017**, *53*, 2731–2744. [CrossRef]
21. Alic, E.; Cermik, O.; Tokgoz, N.; Kaska, O. Optimization of the Pool Boiling Heat Transfer in the Region of the Isolated Bubbles using the ABC Algorithm. *J. Appl. Fluid Mech.* **2019**, *12*, 1241–1248.
22. Benjamin, R.J.; Balakrishnan, A.R. Nucleate pool boiling heat transfer of pure liquids at low to moderate heat fluxes. *Int. J. Heat Mass Transf.* **1975**, *39*, 2495–2504. [CrossRef]

23. Sateesh, G.; Das, S.K.; Balakrishnan, A.R. Analysis of pool boiling heat transfer: Effect of bubbles sliding on the heating surface. *Int. J. Heat Mass Transf.* **2005**, *48*, 1543–1553. [CrossRef]

24. Churchill, S.W.; Chu, H.H.S. Correlating equations for laminar and turbulent free convection from a vertical plate. *Int. J. Heat Mass Transf.* **1975**, *18*, 1323–1329. [CrossRef]

25. Kim, Y.I.; Baek, W.-P.; Chang, S.H. Critical heat flux under flow oscillation of water at low-pressure, low-flow conditions. *Nucl. Eng. Des.* **1999**, *193*, 131–143. [CrossRef]

26. Kim, H.C.; Baek, W.P.; Chang, S.H. Critical heat flux of water in vertical round tubes at low pressure and low flow conditions. *Nucl. Eng. Des.* **2000**, *199*, 49–73. [CrossRef]

27. He, M.; Lee, Y. Application of machine learning for prediction of critical heat flux: Support vector machine for data-driven CHF look-up table construction based on sparingly distributed training data points. *Nucl. Eng. Des.* **2018**, *338*, 189–198. [CrossRef]

28. Cai, J. Applying support vector machine to predict the critical heat flux in concentric-tube open thermosiphon. *Ann. Nucl. Energy* **2012**, *43*, 114–122. [CrossRef]

29. Gandhi, A.B.; Joshi, J.B. Estimation of heat transfer coefficient in bubble column reactors using support vector regression. *Chem. Eng. J.* **2010**, *160*, 302–310. [CrossRef]

30. Nafey, A.S. Neural network based correlation for critical heat flux in steam-water flows in pipes. *Int. J. Therm. Sci.* **2009**, *48*, 2264–2270. [CrossRef]

31. Najafi, H.; Woodbury, K.A. Online heat flux estimation using artificial neural network as a digital filter approach. *Int. J. Heat Mass Transf.* **2015**, *91*, 808–817. [CrossRef]

32. Balcilar, M.; Dalkilic, A.S.; Wongwises, S. Artificial neural network techniques for the determination of condensation heat transfer characteristics during downward annular flow of R134a inside a vertical smooth tube. *Int. Commun. Heat Mass Transf.* **2011**, *38*, 75–84. [CrossRef]

33. Parveen, N.; Zaidi, S.; Danish, M. Development and Analyses of Artificial Intelligence (AI)-Based Models for the Flow Boiling Heat Transfer Coefficient of R600a in a Mini-Channel. *ChemEngineering* **2018**, *2*, 27. [CrossRef]

34. Haupt, R.L.; Haupt, S.E. *Practical Genetic Algorithms*; John Wiley and Sons: Hoboken, NJ, USA, 2004.

35. Karaboga, D.; Basturk, B. A powerful and efficient algorithm for numerical function optimization: Artificial bee colony (ABC) algorithm. *J. Glob. Optim.* **2007**, *39*, 459–471. [CrossRef]

36. Karaboga, D.; Basturk, B. On the performance of artificial bee colony (ABC) algorithm. *Appl. Soft Comput.* **2008**, *8*, 687–697. [CrossRef]

37. Vapnik, V. *The Nature of Statistical Learning Theory*; Springer Science & Business Media: New York, NY, USA, 2013.

38. Schölkopf, B.; Burges, C.J.C.; Smola, A.J. Introduction to support vector learning. In *Advances in Kernel Methods*; MIT Press: Cambridge, MA, USA, 1999; pp. 1–15.

39. Karal, Ö. Compression of ECG data by support vector regression method. *J. Fac. Eng. Arch. Gazi Univ.* **2018**, *1*, 743–756.

40. Das, M.; Akpinar, E. Investigation of Pear Drying Performance by Different Methods and Regression of Convective Heat Transfer Coefficient with Support Vector Machine. *Appl. Sci.* **2018**, *8*, 215. [CrossRef]

41. Guo, W.; Yang, Y.; Zhou, Y.; Tan, Y. Influence Area of Overlap Singularity in Multilayer Perceptrons. *IEEE Access* **2018**, *6*, 60214–60223. [CrossRef]

42. Sun, H.; Hu, X.; Zhang, Y. Attribute Selection Based on Constraint Gain and Depth Optimal for a Decision Tree. *Entropy* **2019**, *21*, 198. [CrossRef]

43. Krishnamoorthy, D.; Foss, B.; Skogestad, S. Real-time optimization under uncertainty applied to a gas lifted well network. *Processes* **2016**, *4*, 52. [CrossRef]

44. Freund, Y.; Mason, L. The alternating decision tree learning algorithm. In Proceedings of the Sixteenth International Conference on Machine Learning (ICML '99), Bled, Slovenia, 27–30 June 1999; Volume 99, pp. 124–133.

45. Frank, E.; Mayo, M.; Kramer, S. Alternating model trees. In Proceedings of the 30th Annual ACM Symposium on Applied Computing, Salamanca, Spain, 13–17 April 2015; pp. 871–878.

46. Pal, M.; Mather, P.M. An assessment of the effectiveness of decision tree methods for land cover classification. *Remote Sens. Environ.* **2003**, *86*, 554–565. [CrossRef]

47. Kavzoglu, T.; Colkesen, I. Classification of Satellite Images Using Decision Trees. *Electron. J. Map Technol.* **2010**, *2*, 36–45.

48. Wang, L.; Zhou, X. Detection of Congestive Heart Failure Based on LSTM-Based Deep Network via Short-Term RR Intervals. *Sensors* **2019**, *19*, 1502. [CrossRef]
49. Jabbar, S.A.; Sultan, A.J.; Maabad, H.A. Prediction of Heat Transfer Coefficient of Pool Boiling Using Back propagation Neural Network. *Eng. Technol. J.* **2012**, *30*, 1293–1305.
50. Sayahi, T.; Tatar, A.; Bahrami, M. A RBF model for predicting the pool boiling behavior of nanofluids over a horizontal rod heater. *Int. J. Therm. Sci.* **2016**, *99*, 180–194. [CrossRef]
51. Hakeem, M.A.; Kamil, M.; Asif, M. Prediction of boiling heat transfer coefficients in pool boiling of liquids using artificial neural network. *J. Sci. Ind. Res.* **2014**, *73*, 536–540.

Article

Enhancement Study of Ice Storage Performance in Circular Tank with Finned Tube

Hua Zhou [1], Mengting Chen [2], Xiaotian Han [2], Peng Cao [2], Feng Yao [1,]* and Liangyu Wu [2,]*

[1] Jiangsu Key Laboratory of Micro and Nano Heat Fluid Flow Technology and Energy Application, School of Environmental Science and Engineering, Suzhou University of Science and Technology, Suzhou 215009, China; hzhou@microflows.net

[2] School of Hydraulic, Energy and Power Engineering, Yangzhou University, Yangzhou 225127, China; mtchen@microflows.net (M.C.); xthan@microflows.net (X.H.); pcao@microflows.net (P.C.)

* Correspondence: yaofeng@usts.edu.cn (F.Y.); lywu@yzu.edu.cn (L.W.); Tel.: +86-514-8797-1315 (L.W.)

Received: 1 April 2019; Accepted: 2 May 2019; Published: 7 May 2019

Abstract: Combined experimental and numerical studies are conducted to study ice storage performance of an ice storage tank with finned tube. Axially arranged fins on the refrigerant tube are applied to enhance the solidification. The evolution of the solid–liquid interface and the variation of temperature of the typical position is examined. The effect of natural convection is discussed in detail. In addition, the effects of refrigerant temperature and initial water temperature on the ice storage performance are analyzed. The results indicate that the ice storage performance is enhanced by the metal fins remarkably. The defection of poor heat transfer after ice is formed can be solved by applying fins in ice storage devices. Natural convection leads to unnecessary mixing of water with different temperatures, lessening the cooling energy stored and acting as a disadvantage during solidification. Decreasing the refrigerant temperature and initial water temperature is beneficial for ice storage.

Keywords: ice storage; finned tube; natural convection; visualization experiment; numerical simulation

---

1. Introduction

Latent heat storage through phase change materials is of great interest in a wide range of technical applications, including electronics cooling [1,2], microfluidics system [3], building energy conservation [4], biomedical engineering [5], space thermal control [6,7], solar energy utilization [8], and chemical process [9,10]. As a typical application, the ice storage technique stores cooling capacity through the solidification of water during the valley electricity demand period and releases the cooling capacity by melting the ice during the peak electricity demand period. Peak load shifting of the electrical grid [11,12] can be achieved through ice storage with low cost, which is of significance in maintaining the steady operation of the electrical grid. Hence, extensive efforts have been devoted to the study of ice storage processes. Conventionally, ice is formed outside a tube or coiler in which refrigerating medium is running. However, thermal resistance between the refrigerating medium and water increases with the thickness of ice due to the low heat transfer coefficient of ice. This poor heat transfer characteristic hinders the widespread use of the ice storage technique. Hence, a number of previous investigations [13,14] have been conducted to improve heat transfer performance, in which methods have included adding extensional structures [15], metal foam [16,17], rough surface structures or particles with high thermal conductivity [18].

Most attempts to explore the natural convection during ice storage processes with a finned tube are based on experimental results [19,20]. Jannesari and Abdollahi [21] studied the effect of a thin ring and annular fin on the improvement of ice formation. The ice formation rate was increased by 15% through the utilization of fins. Ismail and Lino [22] investigated the enhancement of heat transfer in phase change materials (PCM) with a radial finned tube. It was found that the temperature of the refrigerating medium affected the solidification process strongly. Further, the increment in the fin diameter accelerated the velocity at the interface, thus reducing solidification time. Rozenfeld et al. [23] designed a double-pipe heat storage device with helical fins, which shortens the melting time by up to approximately 66%. Four types of solidification of pure water, depending on the strength of natural convection, are reported by Kumar et al. [24]. The enhancement performance of circular and longitudinal fins outside a sleeve-tube is studied by Agyenim et al. [25]. Longitudinal fins proved to be better than circular fins.

Numerical approaches have also been conducted to investigate ice storage processes [26–28]. However, in most research, the flow of the liquid PCM or water was neglected and the effect of natural convection was not taken into account. Jia et al. [29] studied the cold storage of a spherical PCM capsule with fins using the enthalpy method. Ezan and Kalfa [30] numerically investigated the natural convection heat transfer of freezing water inside a square cavity without fins. However, numerical research on the natural convection during solidification in ice storage devices with a finned tube is still lacking [31].

In summary, fluid flow affects heat transfer during solidification, especially at the solid–liquid interface. Besides, the variation of the density of water is non-linear with temperature, which leads to more complicated behavior during solidification. The coupled effect of fins and natural convection on ice storage performance is of significance. Therefore, this article experimentally and numerically investigates the enhancement of ice storage performance in a circular tank by an axially finned tube, in an effort to identify the role of natural convection on ice storage performance. The evolution of the solid–liquid interface and the temperature variation along both the radial and circumferential direction in the water and in the fins, are analyzed. The effect of natural convection on the ice storage process is discussed in particular, based on detailed transit temperature distribution and ice storage ratio. In addition, the effects of temperature of the refrigerants and the initial temperature of the water are discussed.

2. Experimental and Numerical Methods

2.1. Ice Storage Performance Test System

A visualization experimental system was designed as shown in Figure 1, based on which the solid–liquid phase change process during ice storage in a circular tank was studied. The experimental system was comprised of a polymethyl methacrylate ice storage tank, an aluminum finned tube, a flow meter, a constant-temperature bath with working fluid of alcohol and glycol water solution as refrigerant, a light source, a data acquisition instrument (recording the temperature inside the ice storage tank), a computer and a CCD camera monitoring the solid–liquid phase change. The ice storage tank, with a height of 145 mm and inner diameter of 170 mm, was wrapped in foamed polyethylene and tin foil for thermal insulation (Figure 2). The aluminum tube, with eight 50 mm-length fins, was assembled inside the ice storage tank so that the working fluid was flowing inside the tube (Figure 3). During the experiment, the tank was filled with water. In order to measure the temperature of different locations inside the tank during the solid–liquid phase change process, thermocouples were arranged in the water and the fin of the tube through the preset holes, as shown in Figure 4. There were three sets of temperature measurement points in the water, and four sets of temperature measurement points in the fin of the tube. All of the thermocouples were at the same cross section. During the experiment, the temperatures of the refrigerant were −5 °C, −10 °C and −15 °C, and the flow rate was 30 mL/min, with initial water temperatures of 4 °C, 10 °C and 17 °C, respectively. The freezing temperature of glycol water solution is −53 °C while the specific heat is 2.4 kJ/(kg·°C).

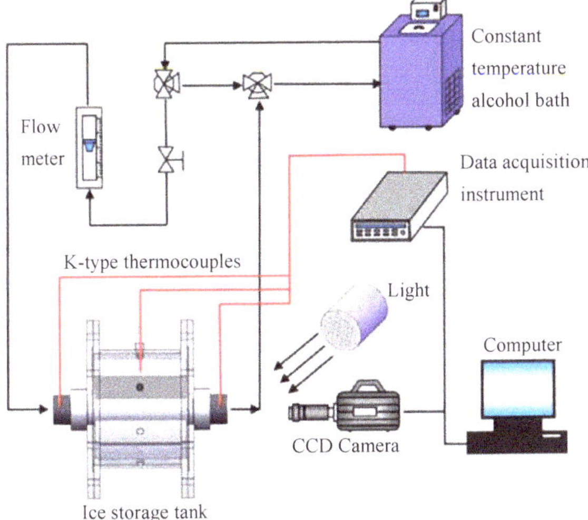

Figure 1. Schematic of the ice storage performance test system.

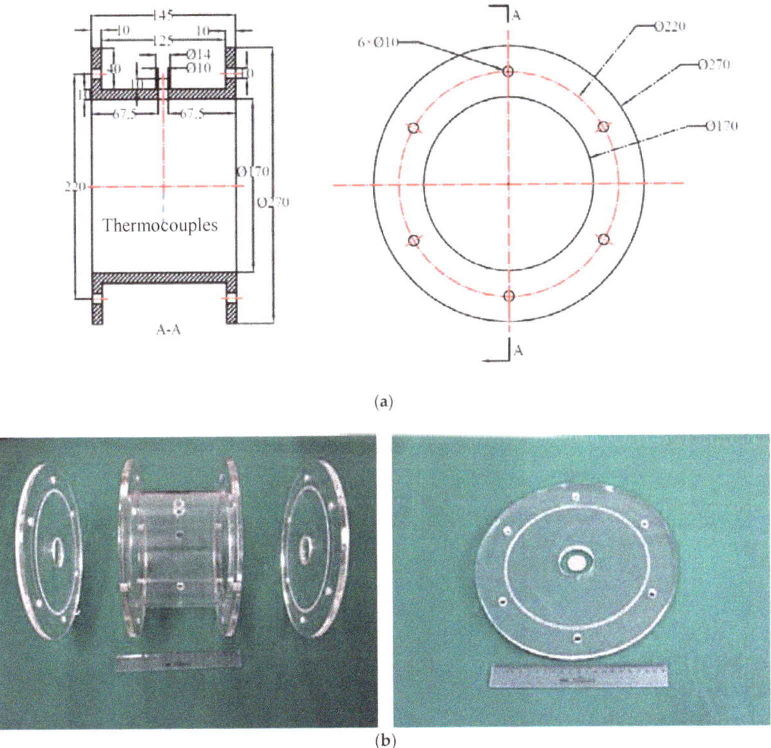

Figure 2. The ice storage tank: (**a**) Schematic; (**b**) photo.

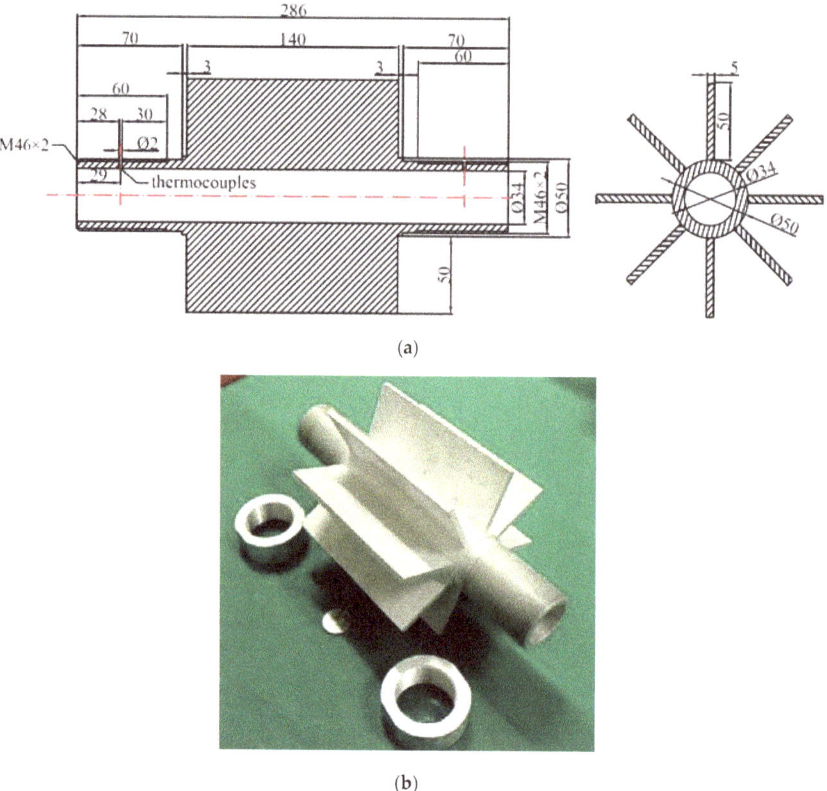

Figure 3. The aluminum tube with fins: (**a**) Schematic; (**b**) photo.

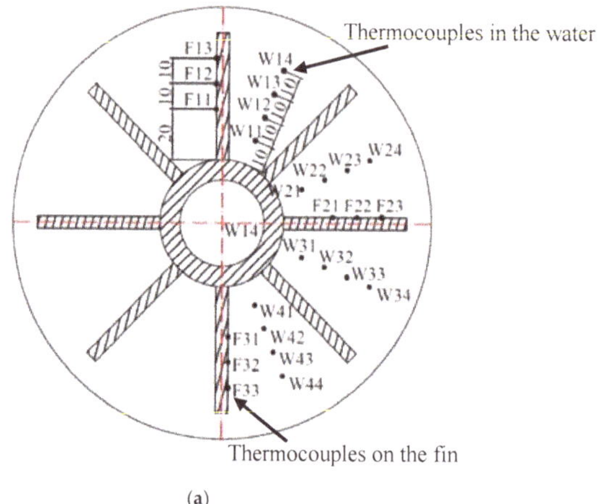

(**a**)

Figure 4. *Cont.*

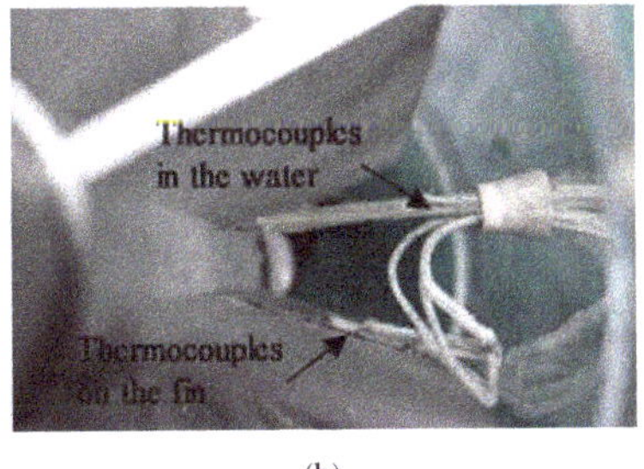

(**b**)

Figure 4. Thermocouple arrangement in the ice storage tank: (**a**) Four sets of thermocouples in the water and three sets of thermocouples on the fin; the distance between the thermocouples in each set is 10 mm; for thermocouples in water, the distance between the innermost thermocouple and the tube is 10 mm; For the thermocouples on the fin, the distance between the innermost thermocouple and the tube is 20 mm; (**b**) photo of thermocouples in the water and on the fin.

2.2. Numerical Methods

To study the enhancement of ice storage performance by arranging external fines on the tube, a transit numerical model, taking natural convection into consideration, was developed based on the enthalpy–porosity method. To simplify the simulation, the following assumptions were made:

1. Water is isotropic in both liquid and solid state, and the physical parameters of water are constant, except density;
2. For the tube with fins, which is also isotropic, the values of the physical parameters are common;
3. The effects of the volume and the supercooling during freezing are neglected, and the phase transition temperature is fixed;
4. There is only laminar flow inside the tank;
5. The fluid and solid are in local heat balance.

Instead of tracking the solid–liquid interface explicitly, a liquid fraction β was introduced to describe the distribution of the liquid and solid as

$$\beta = \begin{cases} 1 & T \geq T_s \\ 0 & T < T_s \end{cases} \tag{1}$$

where T_s is the solidification temperature of water. The thermal capacity and thermal conductivity throughout the entire computational domain were calculated as

$$c = c_s + \beta(c_l - c_s) \tag{2}$$

$$\lambda = \lambda_s + \beta(\lambda_l - \lambda_s) \tag{3}$$

where the subscript s denotes solid and subscript l denotes liquid.

The mass conservation equation is expressed as

$$\frac{\partial \rho}{\partial t} + \frac{\partial(\rho u)}{\partial x} + \frac{\partial(\rho v)}{\partial y} = 0 \tag{4}$$

where ρ is the density, t is the flow time, u is the velocity component along the x-axis and v is the velocity component along the y-axis.

The momentum equation along the x-axis is

$$\frac{\partial(\rho u)}{\partial t} + u\frac{\partial(\rho u)}{\partial x} + v\frac{\partial(\rho u)}{\partial y} = \frac{\partial}{\partial x}(u\frac{\partial u}{\partial x}) + \frac{\partial}{\partial y}(u\frac{\partial u}{\partial y}) - \frac{\partial p}{\partial x} + S_u \tag{5}$$

The momentum equation along the y-axis is

$$\frac{\partial(\rho v)}{\partial t} + u\frac{\partial(\rho v)}{\partial x} + v\frac{\partial(\rho v)}{\partial y} = \frac{\partial}{\partial x}(u\frac{\partial v}{\partial x}) + \frac{\partial}{\partial y}(u\frac{\partial v}{\partial y}) - \frac{\partial p}{\partial x} + S_v \tag{6}$$

where p is the pressure. The source terms in Equations (5) and (6) are

$$S_u = \frac{(1-\beta)^2}{\beta^3 + \xi} A_{mush} u \tag{7}$$

$$S_v = \frac{(1-\beta)^2}{\beta^3 + \xi} A_{mush} v + \frac{\rho g \delta (h - h_{ref})}{c} \tag{8}$$

where ξ is a computational constant that prevents 0 from occurring in the denominator, $A_{mush} = 100{,}000$ is the constant of the mushy zone, g is the gravitational acceleration, δ is the coefficient of thermal expansion, h is the enthalpy, h_{ref} is the reference enthalpy and c is the thermal capacity.

The energy equation is

$$S_h = \frac{\partial(\rho \Delta H)}{\partial t} \tag{9}$$

where the latent heat is calculated as

$$\Delta H = \begin{cases} L, & T \geq T_m \\ 0, & T < T_m \end{cases} \tag{10}$$

where $L = 333{,}400$ J/kg is the latent heat of water.

Natural convection is considered in this work, while the density of water is set to vary with temperature. Since the density of water reaches maximum at 4 °C, a non-linear function was adopted as [32–35]

$$\rho = \rho_{max}(1 - \omega |T - T_{max}|^q) \tag{11}$$

where $\rho_{max} = 999.972$ kg/m^3, $\omega = 9.297173 \times 10^{-6}$·°C, $q = 1.894816$, $T_{max} = 4.02935$·°C. In the current mathematical model, the ice storage was assumed be a process of ice accumulation with little effect of frosting [36].

There is a thermal equilibrium between the water and the fins, which is given by

$$T_{fin,\Omega} = T_{w,\Omega} \tag{12}$$

$$-\lambda_{fin}\left(\frac{\partial T_{fin}}{\partial n}\bigg|_{\Omega}\right) = -\lambda_w\left(\frac{\partial T_w}{\partial n}\bigg|_{\Omega}\right) \tag{13}$$

where the subscript *fin* denotes the fins of the tube, subscript w denotes water, subscript Ω is the boundaries between the water and the fins, and n represents the normal direction of heat flux.

3. Results and Discussion

Based on the aforementioned visualization experiment and numerical model, both experimental and numerical studies on the ice formation in circular tank on the cooling tube with axially arranged fins were conducted, in an effort to clarify the involved natural convection and morphology development. In the current work, we performed repetitive experiments six times to attain reliable results.

3.1. Solidification Process

To examine the evolution of the liquid–solid interface in the ice storage area, the morphologies of the ice during the ice storage process are shown in Figure 5a. The temperature of the refrigerants in this section was −10 °C and the initial temperature of the water was 10 °C. The interface reconstructed from numerical results is shown in Figure 5b (considering the effect of natural convection) and Figure 5d (neglecting natural convection). All the values used in the simulation were the same as the values in the experiments. The temperature of the inner surface of the tube was set to be constant at −10 °C and the initial temperature of the water was set to be 10 °C. The surface of the tank is set to be adiabatic. The initial pressure inside the tank was the same as the environment and the gauge pressure is 0 Pa. To further understand the heat transfer characteristics during solidification, the corresponding temperature contours and velocity vectors from numerical simulation are presented in Figure 5c,e. η is the relative ice storage ratio defined as the quality of ice divided by the total quality of water before solidification.

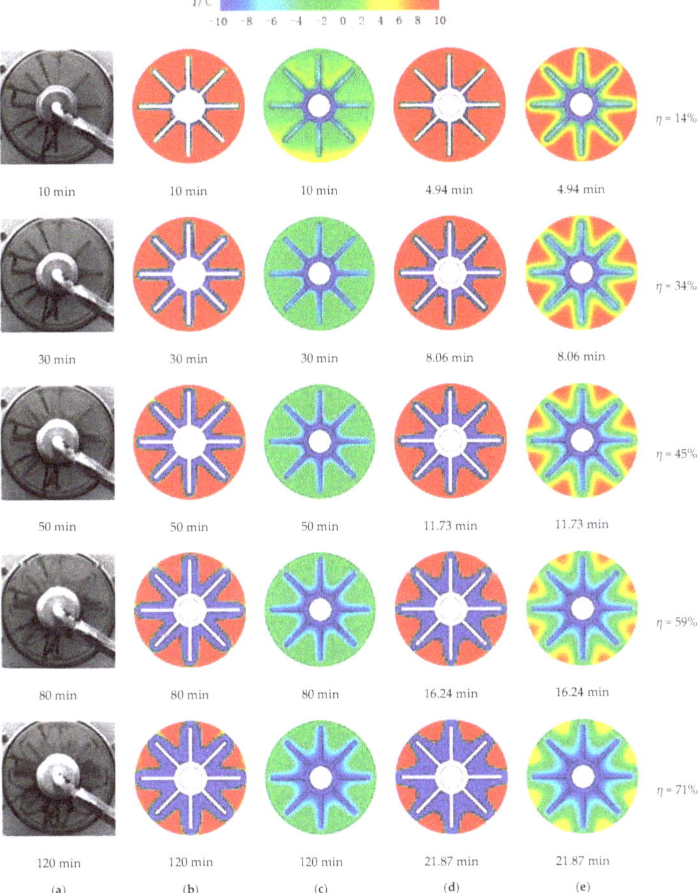

Figure 5. Time evolution of the liquid–solid interface during the ice storage process: (**a**) Experimental snapshots; (**b**) liquid fraction reconstructed from numerical simulation with natural convection and (**c**) the corresponding temperature contours; (**d**) liquid fraction reconstructed from numerical simulation without natural convection and (**e**) the corresponding temperature contours. The relative ice storage ratio η in (**a**,**b**) with natural convection is the same as that in (**c**,**d**) without natural convection.

Evidently, the contour of the ice develops from the inner to the outer region of the tank along the fins and the shape of the ice layer resembles the shape of the fins. As can be seen from both the experimental snapshot and numerical reconstruction figures, the ice is formed on both surfaces of the tube fins at t = 10 min. This is attributed to the fast cooling effect brought by the fins and the temperature of the fins are much lower than the bulk region of water, as shown in Figure 5c. At the beginning of solidification, the thickness of the ice on the fins decreases slightly along the lengthwise direction of the fins due to the temperature gradient on the fins. The relative ice storage ratio increases with time, as shown in Figure 6. As solidification continues, the ice at the bottom of the fins connects at the adjacent fins, meaning that the ice at the bottom of the fins is thicker under the condition of no natural convection. When the cooling energy is distributed more evenly in the water, the ice layer has equivalent thickness along the fins under the influence of natural convection. The rate of solidification is at its maximum at the beginning of solidification, which can be ascribed to the high heat transfer temperature difference and low thermal resistance during this period. As the ice grows thicker and the water in the tank cools, the heat transfer temperature difference drops and the thermal resistance increases, leading to slowing down in the solidification process, as shown in Figure 6.

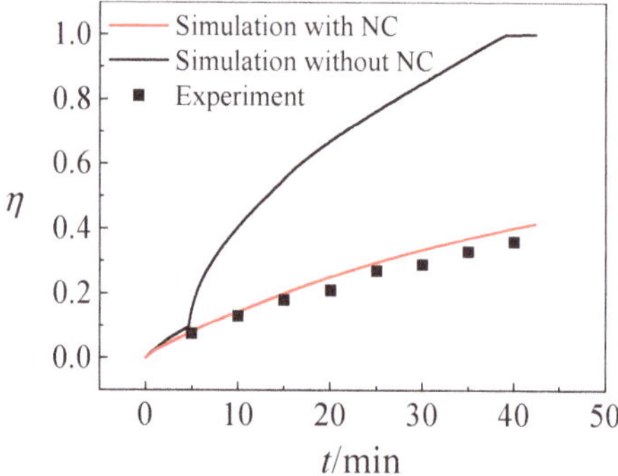

Figure 6. Variation of ice storage ratio with time (NC: nature convection).

As soon as the temperature gradient develops in the water, natural convection occurs. It can be seen from the comparison between Figure 5b,d that the contours of the ice differ and solidification requires less time if there is no natural convection. The variation of ice thickness along the lengthwise direction of the fins is less obvious under the condition of natural convection due to the transfer of cooling energy by the flow of the water and because the temperature inside the tank is no longer longitudinally symmetrical. The temperature of water at the bottom of the sub-regions is also higher than that at the top. Consequently, solidification is faster at the top of fins than at the bottom, especially in the early stage. In addition, Figure 7 shows the natural convection in the tank. It can be seen that tank is divided into several sub-regions by the fins and some vortexes are observed in each sub-region with different profiles. Note that the number of vortexes decreases with time, suggesting that natural convection is suppressed by the increasing confinement in each sub-region with growing ice. In addition, vortexes are asymmetrical between the top and bottom of the tank due to uneven pressure distribution in the sub-regions caused by gravity. The numerical results indicate that, to store the same amount of ice, only about 20% of time is required in the system without natural convection. Decouple time is required for all the water to solidify under the influence of natural convection. This is mainly a result of the mixing of cold water around the fins and the warm water away from the fins. In particular, a large portion of the cooling energy is consumed to cool the bulk of the water through sensible heat at the beginning of the solidification. Only a small amount of cooling energy can be used to overcome the latent heat of the water around the fins and accomplish the solidification process. This can be verified by the temperature distribution inside the tank, as illustrated in Figure 5c,e. In the system without natural convection cooling energy can only be transferred through heat conduction.

Hence the water near the shell of the tank was still warm even when the solidification process was over halfway complete (t = 16.24 min in Figure 5e). Water is cooler in the system with natural convection possessing smaller heat transfer difference.

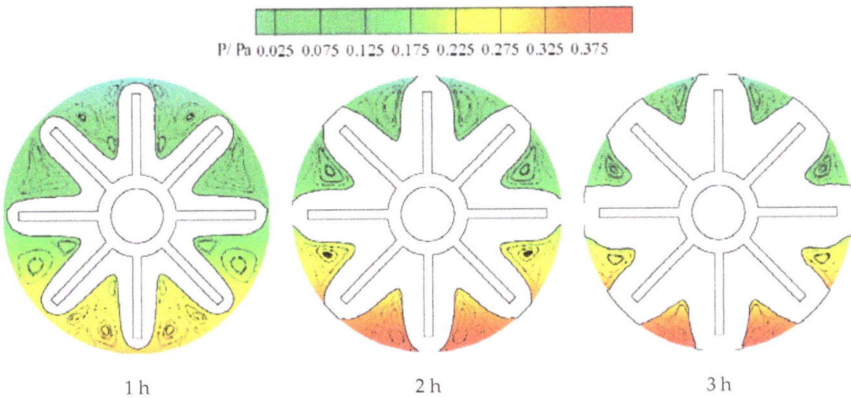

Figure 7. Natural convection flow pattern during the solidification.

In conclusion, natural convection causes the mixing of water and brings a disadvantage to the ice storage technique, especially in the system using materials with low viscosity. To reduce natural convection, materials with high viscosity and small density variation with temperature are recommended and fin structures can be designed to hinder the fluid flow.

To further understand the heat transfer affected by natural convection along the radial direction during solidification, experimental measurements of temperature variation of the points arranged radially in the tank are plotted in Figure 8. Note that the indexes of temperature points in Figures 8–12 are given in Figure 4 and can also be seen in the simplified schematic added in each inset. The points in Figure 8a,b are symmetric about the x-axis and the points in Figure 8a are higher than the points in Figure 8b. In both two different radial directions, the decrease in the temperature is rapid at the beginning and then gradually slows down. The temperature of the points closer to the tube such as point W21 is lower than those away from the tube such as point W24. However, due to the mixing behavior of fluids, temperature differences between the points at the outer rings (such as points W22, W23 and W24) are less obvious. Temperature fluctuations are observed in all points measured, resulting from the vortexes driven by natural convection. Besides, the temperature difference of the measuring points in Figure 8a is smaller than that in Figure 8b, indicating that heat transfer is intense in the upper part of the ice storage tank.

Caused by natural convection, water also flows in the circumferential direction during heat transfer. The evolution of the temperature of circumferentially arranged points with different distance from the tube is shown in Figure 9. Variation of the temperature for all the measuring points follows a similar tendency. However, the temperature of the points in the upper part of the ice storage tank is higher than those in the lower part due to natural convection. The temperature difference between the points closer to the tube is higher (Figure 9a) than those away from the tube (Figure 9c), indicating that the mixing behavior in the region away from the tube is stronger. This is due to the larger space for natural convection that the fluid flow is less confined.

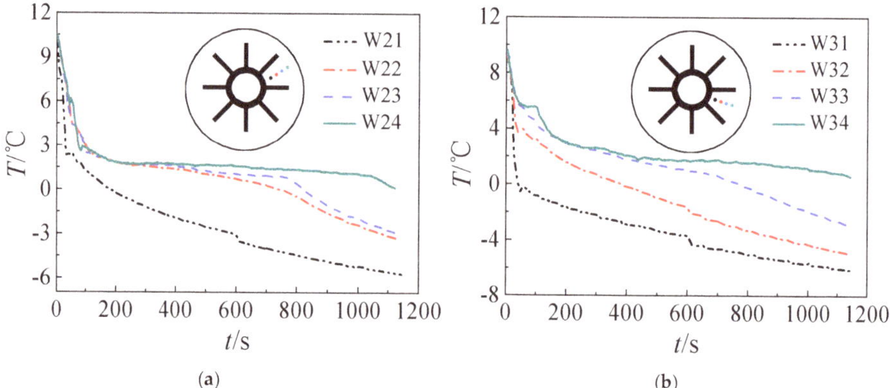

(a)

(b)

Figure 8. Variation of temperatures in the freezing area along the radial direction: (**a**) Points above *x*-axis; (**b**) points below *x*-axis.

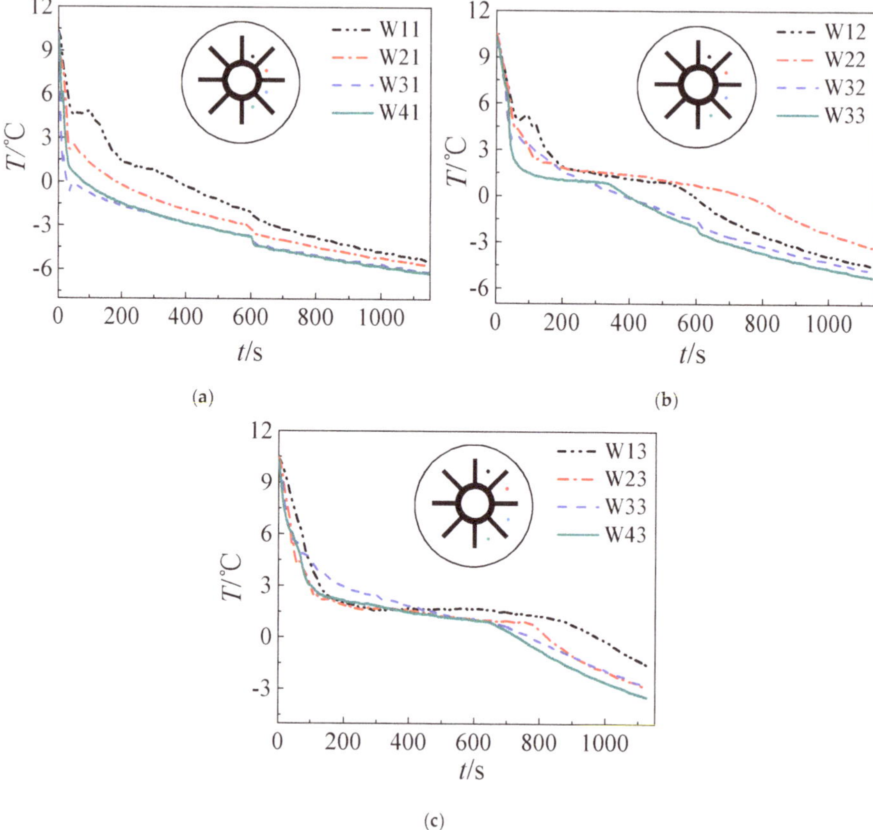

(a)

(b)

(c)

Figure 9. Time evolution of the temperatures in the freezing area along the circumferential direction: (**a**) Distance to the tube is 10 mm; (**b**) distance to the tube is 20 mm; (**c**) distance to the tube is 30 mm.

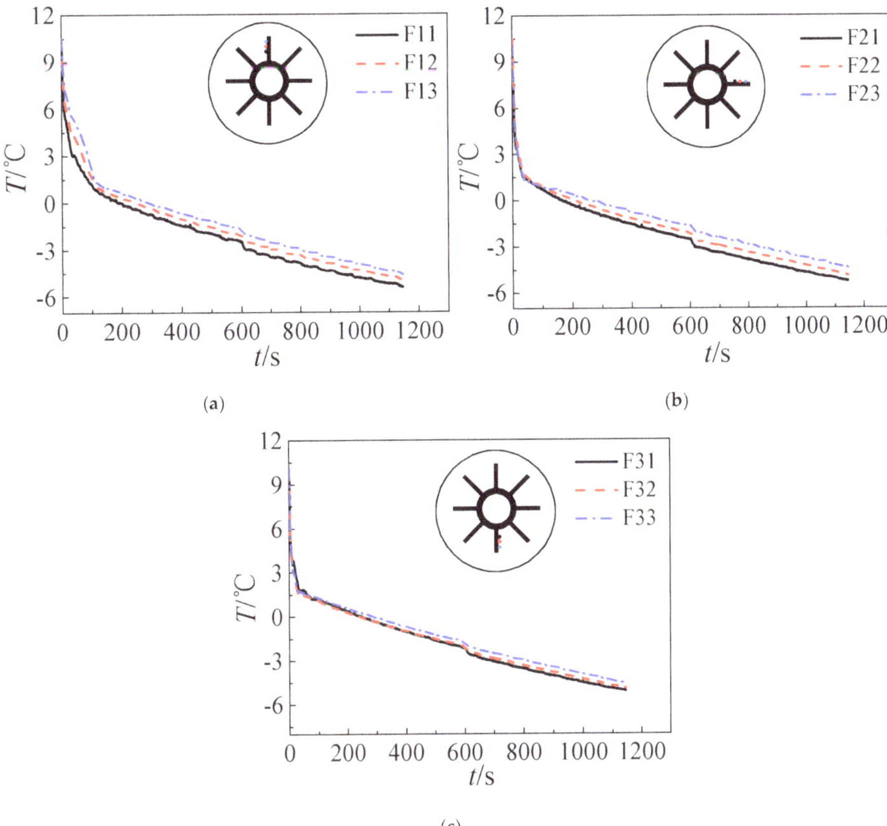

Figure 10. Time evolution of the temperatures of the fins: (**a**) uppermost vertical fins; (**b**) horizontal fins; (**c**) nethermost vertical fins.

Temperature distribution in the fins is similar in all three fins examined, as shown in Figure 10. At the beginning, the temperature drops rapidly due to the high temperature difference and heat transfer rate. This decrease slows down when solidification begins. The temperature of points close to the tube is the lowest, but the temperature difference of the fin along the radial direction is much smaller than that of the water due to the high thermal conductivity of the aluminum. Comparing Figures 8 and 10, the temperature difference between the fins and water is still high at the top of the fins, indicating that heat transfer can be enhanced greatly by fins. Also, a slight temperature fluctuation in the fins is observed and this is caused by the heat transfer with water whose temperature is affected by natural convection.

3.2. Effect of the Temperature of the Refrigerants

Table 1 shows the time evolution of liquid–solid interface during the ice storage processes with different temperature of the refrigerants. The decrease in the temperature of the refrigerants evidently accelerates the ice storage process. The volume of ice with the lower temperature of the refrigerant is greater. To analyze the effect of the refrigerant temperature quantitatively, Figure 11 gives the variation of the temperature in the freezing area. In all three conditions, the temperature drops rapidly due the large temperature difference and the enhancement of the heat transfer by the fins is intense during the pre-solidification phase. Following this, the ice gets thicker, which leads to a deterioration of heat transfer. Besides, a portion of the cooling energy is stored in the ice in the form of latent heat. The temperature decrease slows down soon after for all three refrigerant temperatures. During the solidification process, the temperature is stable. However, the solidification process is shortened by a lower refrigerant temperature. The temperature of the ice decreases with the refrigerant temperature, indicating that more cooling energy is stored in the form of sensible heat in ice in the system with low refrigerant temperature during the post-solidification phase. At the end of the ice storage process, temperature fluctuations can be ignored due to the weak natural convection with small liquid fraction.

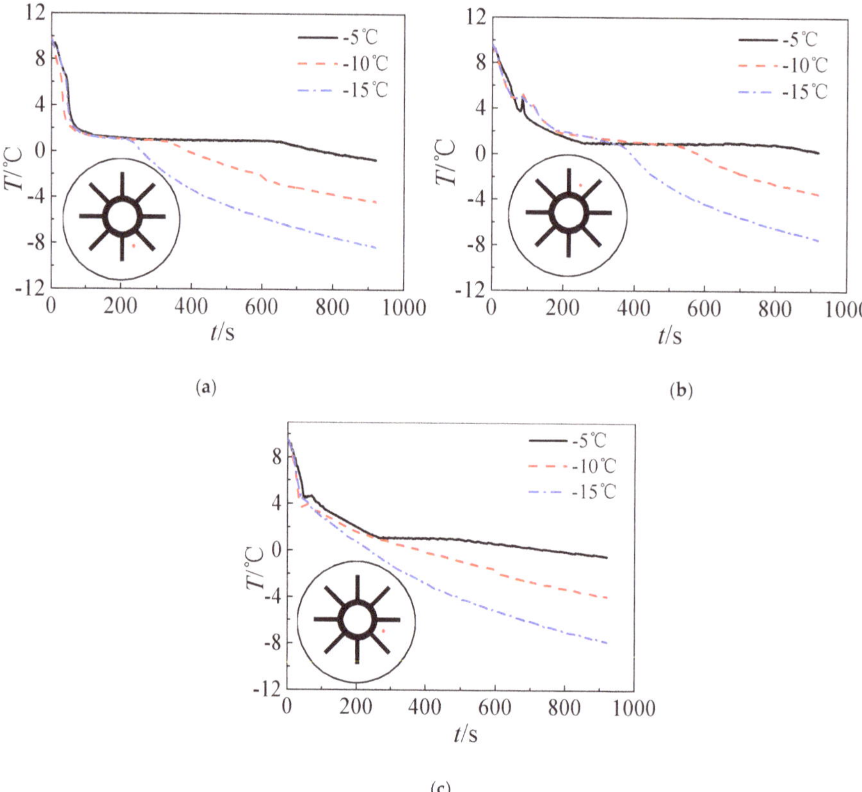

Figure 11. Effect of the refrigerant temperature on the time evolution of the temperatures in the freezing area: (**a**) Point W42; (**b**) Point W12; (**c**) Point W32.

Table 1. Time evolution of liquid–solid interface during the ice storage process with different refrigerant temperatures.

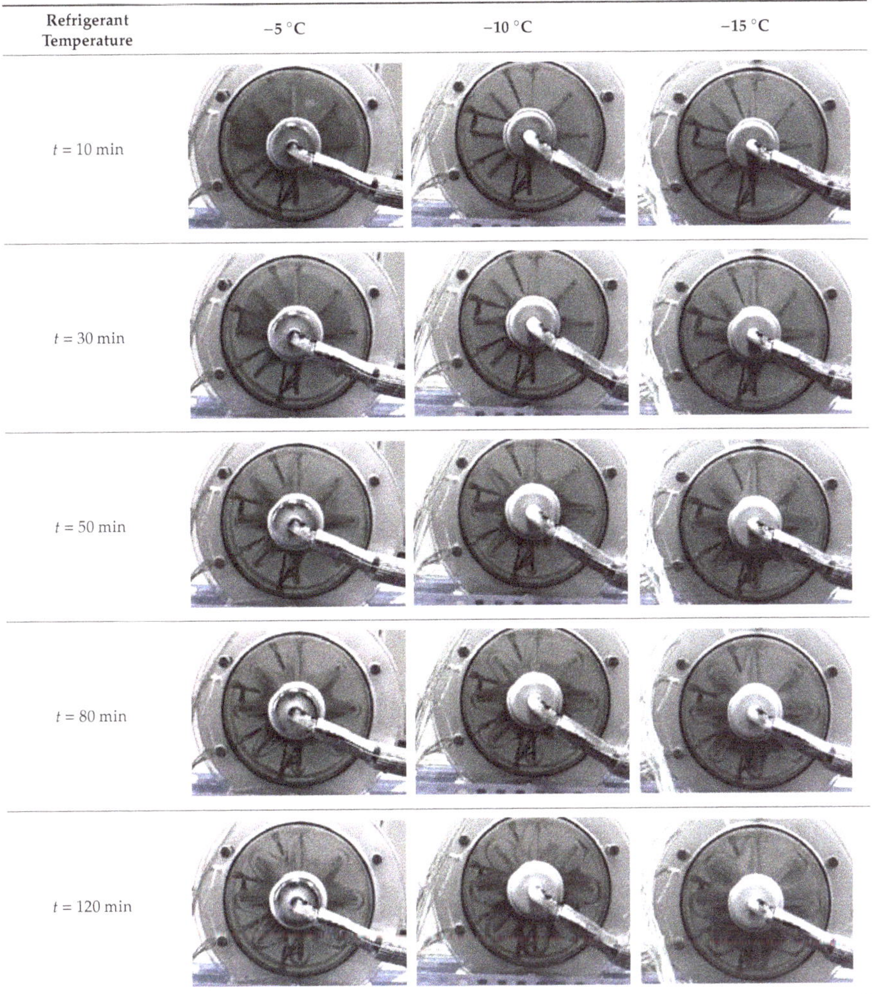

Refrigerant Temperature	−5 °C	−10 °C	−15 °C
t = 10 min			
t = 30 min			
t = 50 min			
t = 80 min			
t = 120 min			

3.3. Effect of the Initial Temperature of the Water

Table 2 shows the time evolution of the liquid–solid interface during the ice storage process with different initial temperatures of water. Comparing the contours of ice in Tables 1 and 2, the influence of the initial temperature of the water is more evident than the effect of the refrigerant temperature in the pre-solidification phase. The volume of ice with a lower initial temperature of the water is much larger, indicating that the cooling of water before solidification consumes a large amount of cooling energy. As shown in Figure 12, the temperature drops rapidly in the system with high initial water temperature due to the large temperature difference between the water and refrigerant. Afterwards, the temperature is relatively stable due to the solidification. The duration of the stable period is similar in all three conditions tested; a slight difference is caused by the different strength of the natural convection.

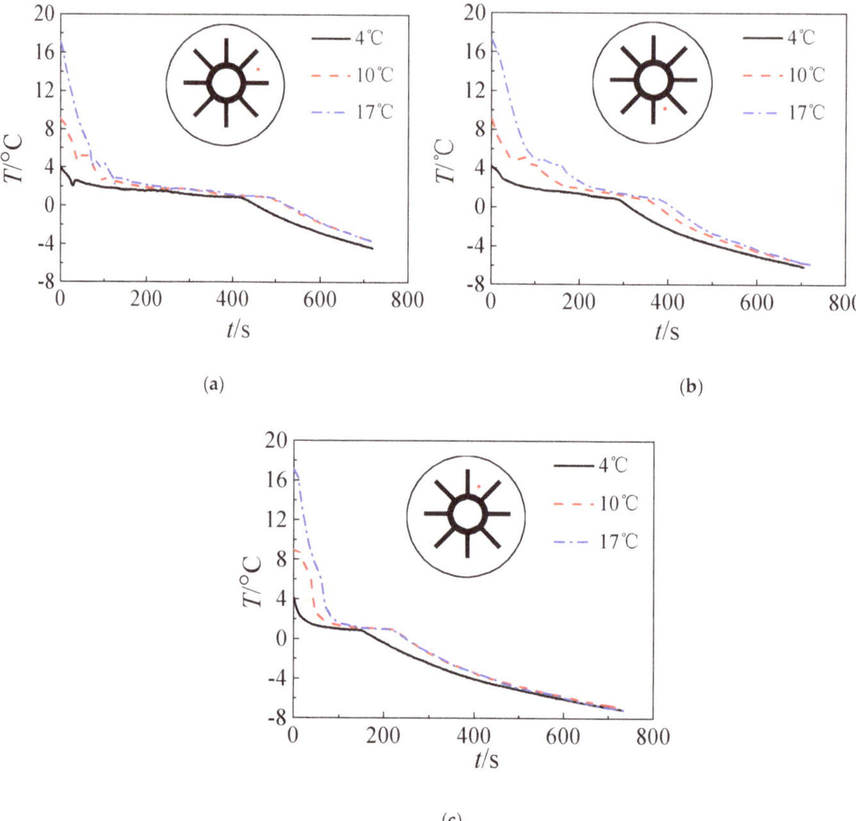

Figure 12. Effect of the initial temperatures of the water on the time evolution of the temperatures in the freezing area: (**a**) Point W23; (**b**) Point W42; (**c**) Point W12.

Table 2. Time evolution of the liquid–solid interface during the ice storage process with different initial water temperatures.

Initial Temperature of the Water	4 °C	10 °C	17 °C
$t = 10$ min			
$t = 30$ min			
$t = 50$ min			
$t = 80$ min			
$t = 120$ min			

Finally, the temperature of the same measuring point is independent with the initial temperature of the water, indicating that the effect of the initial temperature of the water is weak at the post-solidification phase.

4. Conclusions

The combined experimental and numerical studies were conducted to study ice storage performance of an ice storage tank with a finned tube. The evolution of the solid–liquid interface and the temperature variation in both the radial and circumferential directions in the water and in the fins are analyzed. The effect of natural convection on the ice storage process is discussed, with particular focus on the detailed transit of temperature distribution and ice storage ratio. In addition, via the comparison between the performances of the ice storage tank under different working conditions, the effects of the refrigerant temperature and initial water temperature on the ice storage performance are discussed. The main conclusions drawn are as follows:

1. Compared with solidification without natural convection, double the time or more is required to store the same amount of ice under the influence of natural convection. Decouple time is required for the device with natural convection to eventually solidify all the water.

2. Both numerical simulation and experimental results show that the freezing area is larger in the upper part of the tank and the temperature of the upper part is lower due to natural convection. Additionally, the liquid fraction and water area is large at the initial stage, during which natural convection is strong. Water with different temperature mixes that a large amount of cooling energy is consumed in cooling the water. The heat transfer temperature difference is reduced and less cooling energy can be occupied in solidification, indicating that natural convection is disadvantageous in ice storage systems.

3. Variations of refrigerant temperature do not obviously affect the pre-solidification phase. However, the time required by solidification shortens with refrigerant temperature and more cooling energy can be stored in ice in the form of sensible heat in the post-solidification phase.

4. The duration of the pre-solidification phase increases with increasing initial water temperature. However, the influence the initial water temperature on the temperature of the ice weakens with time.

Author Contributions: Investigation, H.Z., X.H., M.C. and P.C.; supervision, L.W. and F.Y.

Funding: This research was funded by National Natural Science Foundation of China (No. 51706194) and Natural Science Foundation of Yangzhou City (No. YZ2017103).

Conflicts of Interest: The authors declare no conflict of interest.

References

1. Kandasamy, R.; Wang, X.-Q.; Mujumdar, A.S. Transient cooling of electronics using phase change material (pcm)-based heat sinks. *Appl. Therm. Eng.* **2008**, *28*, 1047–1057. [CrossRef]

2. Zhang, C.; Yu, F.; Li, X.; Chen, Y. Gravity–capillary evaporation regimes in microgrooves. *AIChE J.* **2019**, *65*, 1119–1125. [CrossRef]

3. Chen, Y.; Liu, X.; Shi, M. Hydrodynamics of double emulsion droplet in shear flow. *Appl. Phys. Lett.* **2013**, *102*, 051609. [CrossRef]

4. Cui, Y.; Xie, J.; Liu, J.; Pan, S. Review of phase change materials integrated in building walls for energy saving. *Procedia Eng.* **2015**, *121*, 763–770. [CrossRef]

5. Chen, Y.P.; Gao, W.; Zhang, C.B.; Zhao, Y.J. Three-dimensional splitting microfluidics. *Lab Chip* **2016**, *16*, 1332–1339. [CrossRef]

6. Fixler, S.Z. Satellite thermal control using phase-change materials. *J. Spacecr. Rockets* **1966**, *3*, 1362–1368. [CrossRef]

7. Liu, X.; Chen, Y.; Shi, M. Dynamic performance analysis on start-up of closed-loop pulsating heat pipes (clphps). *Int. J. Therm. Sci.* **2013**, *65*, 224–233. [CrossRef]

8. Mao, Q.; Chen, H.; Yang, Y. Energy storage performance of a pcm in the solar storage tank. *J. Therm. Sci.* **2019**, *28*, 195–203. [CrossRef]

9. Zhang, C.; Shen, C.; Chen, Y. Experimental study on flow condensation of mixture in a hydrophobic microchannel. *Int. J. Heat Mass Transf.* **2017**, *104*, 1135–1144. [CrossRef]

10. Zhang, C.; Chen, Y.; Wu, R.; Shi, M. Flow boiling in constructal tree-shaped minichannel network. *Int. J. Heat Mass Transf.* **2011**, *54*, 202–209. [CrossRef]

11. Lo, C.-C.; Tsai, S.-H.; Lin, B.-S. Ice storage air-conditioning system simulation with dynamic electricity pricing: A demand response study. *Energies* **2016**, *9*, 113. [CrossRef]

12. Galindo-Luna, Y.R.; Gómez-Arias, E.; Romero, R.J.; Venegas-Reyes, E.; Montiel-González, M.; Unland-Weiss, H.E.K.; Pacheco-Hernández, P.; González-Fernández, A.; Díaz-Salgado, J. Hybrid solar-geothermal energy absorption air-conditioning system operating with naoh-h2o—las tres vírgenes (baja california sur), "la reforma" case. *Energies* **2018**, *11*, 1268. [CrossRef]

13. Deng, Z.; Liu, X.; Zhang, C.; Huang, Y.; Chen, Y. Melting behaviors of pcm in porous metal foam characterized by fractal geometry. *Int. J. Heat Mass Transf.* **2017**, *113*, 1031–1042. [CrossRef]

14. Wang, J.; Sun, L.; Zou, M.; Gao, W.; Liu, C.; Shang, L.; Gu, Z.; Zhao, Y. Bioinspired shape-memory graphene film with tunable wettability. *Sci. Adv.* **2017**, *3*, e1700004. [CrossRef]

15. Zhang, D.H.; Shen, Y.L.; Zhou, Z.P.; Qu, J.; Xu, H.Y.; Cao, W.; Song, D.; Zhang, F.M. Convection heat transfer performance of the fractal tube bank under cross flow. *Fractals* **2018**, *26*, 1850073. [CrossRef]

16. Xiao, B.Q.; Wang, W.; Fan, J.T.; Chen, H.X.; Hu, X.L.; Zhao, D.S.; Zhang, X.; Ren, W. Optimization of the fractal-like architecture of porous fibrous materials related to permeability, diffusivity and thermal conductivity. *Fractals* **2017**, *25*, 9. [CrossRef]

17. Yang, J.; Yang, L.; Xu, C.; Du, X. Experimental study on enhancement of thermal energy storage with phase-change material. *Appl. Energy* **2016**, *169*, 164–176. [CrossRef]

18. Wang, J.; Gao, W.; Zhang, H.; Zou, M.H.; Chen, Y.P.; Zhao, Y.J. Programmable wettability on photocontrolled graphene film. *Sci. Adv.* **2018**, *4*, eaat7392. [CrossRef] [PubMed]

19. Cheng, K.C.; Inaba, H.; Gilpin, R.R. Effects of natural convection on ice formation around an isothermally cooled horizontal cylinder. *J. Heat Transf.* **1988**, *110*, 931–937. [CrossRef]

20. Habeebullah, B.A. An experimental study on ice formation around horizontal long tubes. *Int. J. Refrig.* **2007**, *30*, 789–797. [CrossRef]

21. Jannesari, H.; Abdollahi, N. Experimental and numerical study of thin ring and annular fin effects on improving the ice formation in ice-on-coil thermal storage systems. *Appl. Energy* **2017**, *189*, 369–384. [CrossRef]

22. Ismail, K.A.R.; Lino, F.A.M. Fins and turbulence promoters for heat transfer enhancement in latent heat storage systems. *Exp. Therm. Fluid Sci.* **2011**, *35*, 1010–1018. [CrossRef]

23. Rozenfeld, A.; Kozak, Y.; Rozenfeld, T.; Ziskind, G. Experimental demonstration, modeling and analysis of a novel latent-heat thermal energy storage unit with a helical fin. *Int. J. Heat Mass Transf.* **2017**, *110*, 692–709. [CrossRef]

24. Kumar, V.; Srivastava, A.; Karagadde, S. Real-time observations of density anomaly driven convection and front instability during solidification of water. *J. Heat Trans.* **2018**, *140*, 042503. [CrossRef]

25. Agyenim, F.; Eames, P.; Smyth, M. A comparison of heat transfer enhancement in a medium temperature thermal energy storage heat exchanger using fins. *Sol. Energy* **2009**, *83*, 1509–1520. [CrossRef]

26. Ismail, K.A.; Henriquez, J.; Da Silva, T. A parametric study on ice formation inside a spherical capsule. *Int. J. Therm. Sci.* **2003**, *42*, 881–887. [CrossRef]

27. Drees, K.H.; Braun, J.E. Modeling of area-constrained ice storage tanks. *Hvac R Res.* **1995**, *1*, 143–158.

28. Kousksou, T.; Arid, A.; Majid, J.; Zeraouli, Y. Numerical modeling of double-diffusive convection in ice slurry storage tank. *Int. J. Refrig.* **2010**, *33*, 1550–1558. [CrossRef]

29. Jia, X.; Zhai, X.; Cheng, X. Thermal performance analysis and optimization of a spherical pcm capsule with pin-fins for cold storage. *Appl. Therm. Eng.* **2019**, *148*, 929–938. [CrossRef]

30. Ezan, M.A.; Kalfa, M. Numerical investigation of transient natural convection heat transfer of freezing water in a square cavity. *Int. J. Heat Fluid Flow* **2016**, *61*, 438–448. [CrossRef]

31. Sciacovelli, A.; Gagliardi, F.; Verda, V. Maximization of performance of a pcm latent heat storage system with innovative fins. *Appl. Energy* **2015**, *137*, 707–715. [CrossRef]

32. Voller, V.R.; Prakash, C. A fixed grid numerical modelling methodology for convection-diffusion mushy region phase-change problems. *Int. J. Heat Mass Transf.* **1987**, *30*, 1709–1719. [CrossRef]

33. Brent, A.D.; Voller, V.R.; Reid, K.J. Enthalpy-porosity technique for modeling convection-diffusion phase change: Application to the melting of a pure metal. *Numer. Heat Transf.* **1988**, *13*, 297–318. [CrossRef]

34. Vu, T.V. Three-phase computation of solidification in an open horizontal circular cylinder. *Int. J. Heat Mass Transf.* **2017**, *111*, 398–409. [CrossRef]

35. Fertelli, A.; Günhan, G.; Buyruk, E. Numerical investigation of effect of the position of the cylinder on solidification in a rectangular cavity. *Heat Mass Transf.* **2017**, *53*, 687–704. [CrossRef]

36. Huang, Y.; Khajepour, A.; Bagheri, F.; Bahrami, M. Optimal energy-efficient predictive controllers in automotive air-conditioning/refrigeration systems. *Appl. Energy* **2016**, *184*, 605–618. [CrossRef]

Article

The Effects of Water Friction Loss Calculation on the Thermal Field of the Canned Motor

Yiping Lu [1,*], Azeem Mustafa [2], Mirza Abdullah Rehan [2], Samia Razzaq [2], Shoukat Ali [2], Mughees Shahid [2] and Ahmad Waleed Adnan [2]

1   Department of Mechanical and Power Engineering, Harbin University of Science and Technology, Harbin 150080, China
2   Department of Mechanical Engineering, Pakistan Institute of Engineering and Technology, Multan 66000, Pakistan; azeemmustafa@piet.edu.pk (A.M.); engr.mirzaabdullah58@gmail.com (M.A.R.); samiaa.razzaq@gmail.com (S.R.); shoukatalimugheri@piet.edu.pk (S.A.); mugheesshahid@piet.edu.pk (M.S.); waleedgrewal@gmail.com (A.W.A.)
*   Correspondence: luyiping@hrbust.edu.cn; Tel.: +86-188-0462-1752

Received: 9 March 2019; Accepted: 25 April 2019; Published: 3 May 2019

Abstract: The thermal behavior of a canned motor also depends on the losses and the cooling capability, and these losses cause an increase in the temperature of the stator winding. This paper focuses on the modeling and simulation of the thermal fields of the large canned induction motor by different calculation methods of water friction loss. The values of water friction losses are set as heat sources in the corresponding clearance of water at different positions along the duct and are calculated by the analytical method, loss separation test method, and by assuming the values that may be larger than the experimental results and at zero. Based on Finite volume method (FVM), 3D turbulent flow and heat transfer equations of the canned motor are solve numerically to obtain the temperature distributions of different parts of the motor. The analysis results of water friction loss are compared with the measurements, obtained from the total losses using the loss separation method. The results show that the magnitude of water friction loss within various parts of the motor does not affect the position of peak temperature and the tendency of the temperature distribution of windings. This paper is highly significant for the design of cooling structures of electrical machines.

Keywords: water friction loss; three-dimensional temperature field; numerical simulation; canned motor; computational fluid dynamics (CFD)

---

1. Introduction

The canned motors are currently limited by bearing technologies and the thermal field, preventing a high reliability and long lifetime. The thermal behavior of electric machines can be determined by losses and cooling capability [1,2]. Generally, the losses can be divided into water friction loss, electric loss, including rotor and stator shield loss, stator winding loss, rotor copper bar loss, conical ring loss, iron loss, and eddy-current loss. Electric loss and eddy-current loss can be obtained by the finite element method (FEM) [3]. Due to the characteristics of high thermal density per unit volume of rotor and stator shield and small water clearance in the shield, the water friction loss distribution in canned motors differs from conventional standard motors. The most temperature-sensitive component of the motor is the radial bearings, and their operating temperature should be lower than the alarm value. The effects of losses on the temperature rise of the motor components, such as stator windings, insulations, and radial bearings, is a very important issue [4,5]. Water friction loss calculation methods and thermal analysis, thus, are an important step in the designing of induction motors. Moreover, temperature variations also affect the circulation of cooling fluids and their flow dynamics [6].

Typically, thermal analysis can be carried out using computational fluid dynamics (CFD) [6–8], the heat equation finite element analysis (FEA) [4], or lumped parameter thermal network (LPTN) models [9–11], and the first two are very accurate and deliver proper predictions of the thermal system behavior during the component/device development. Finite volume method (FVM) is one of the CFD methods and has the advantage of predicting the flow in complex regions and obtaining the temperature distribution [8].

With the development of high speed motors and their applications, the research and study of the effect of different losses of motors have become topics of high interest. Luomi et al. [12] studied and proposed a method for loss minimization including air-friction losses and explained that the design is considerably influenced by the air-friction losses, resulting in a small diameter of the rotor. Wang et al. [13] analyzed rotor losses of permanent magnet synchronous machine (PMSM) with experiments and simulations and focused on rotor loss in the no load running condition. Aglen and Andersson [14] theoretically and experimentally studied and calculated different losses of the generator. They utilized a generator model to inspect the temperature distribution and showed that different losses strongly affect the temperature distribution, particularly the temperature of the magnet due to rotor losses. Leakage effects while proposing a design method for bearing were studied by Le Yun et al. [15]. They discussed the effect of leakages in terms of static and dynamic stiffness. Zhang and Kirtely et al. [16] investigated and predicted the power losses including core losses, winding losses, and air-friction losses. The effects of friction loss in electric machines were often neglected in past research; however, in recent years, the friction loss has been considered one of the important factors that might affect the bearing life of the canned motor and account for the percentage of fly wheel loss of the motor due to the increased rotational speeds. As in [12], a method for efficiency optimization of a 500,000-r/min permanent magnet machine has been studied.

There is no study on the friction loss and calculation method of the water-cooled canned motor for the temperature rise of internal components and bearing safety. When the electric loss of the motor is constant, the magnitude and calculation method of friction loss have an influence on the peak temperature of the winding. The research object is a canned motor in this paper; a calculation method is presented to study the effect of friction losses on the thermal field of the motor, and the effect on the peak temperature and temperature variation characteristics of important parts are analyzed. The study has been carried out by computational fluid dynamics (CFD) approaches adopting five friction loss values: theoretical, experimental, assumed 1.1-times experimental, and the CFD module; the last one is the ideal case assuming no friction loss.

2. Physical Model

The studied electrical machine was a vertical squirrel cage motor with a rotational speed of 1500 r/min, 6600 kW (see Figure 1). It had a three-phase, four-pole, short-pitched, and double-layer winding stator whose outside diameter was 1.3 m. The air gap clearance was 4.2 mm; the thickness of the stator and rotor shield casing was 0.46 mm and 0.70 mm, respectively, and the dimensionless ratio between core length and rotor inside diameter was 6.5. It had N-class insulation of the winding with a permitted temperature of 200 °C.

In Figure 1, 1 indicates the outlet, 2 the upper nitrogen end room, 3 the water inlet, 4 the cooling jacket, 5 the stator plate, 6 the upper radial bearing, 7 the stator core, 8 the winding end, 9 the frame, 10 the lower radial bearing, 11 the primary water inlet, 12 the lower fly wheel, 13 the axis, 14 the cage rotor, and 15 the stator can.

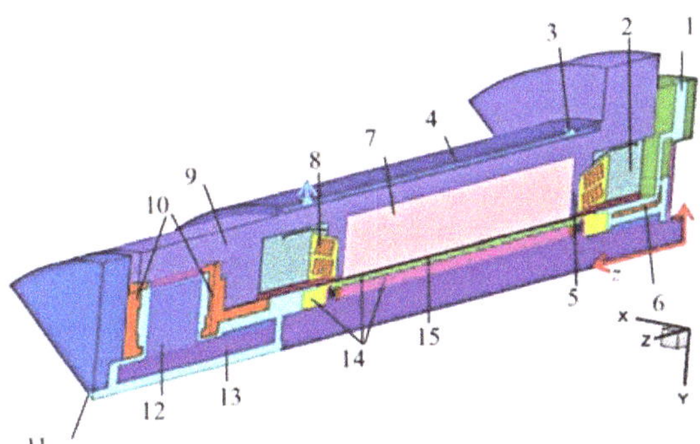

Figure 1. Physical model of the computational domain showing different parts and flow directions.

The temperature of each part can be reduced by the effective internal and external circuit design of the primary cooling water. The secondary cooling water flowed from the upper part of the outer casing (bottom outflow); see Figure 2. The primary water flowing out of the external heat exchanger was drawn through the axial suction hole at the bottom of the frame. Water continued to flow upward to the auxiliary impeller position and was thrown away by the Coriolis force. A part of the cooling water flowed down for the cooling and lubrication of the lower guide bearing and lower fly wheel, then entered into the cycle of cold water confluence and moved downward.

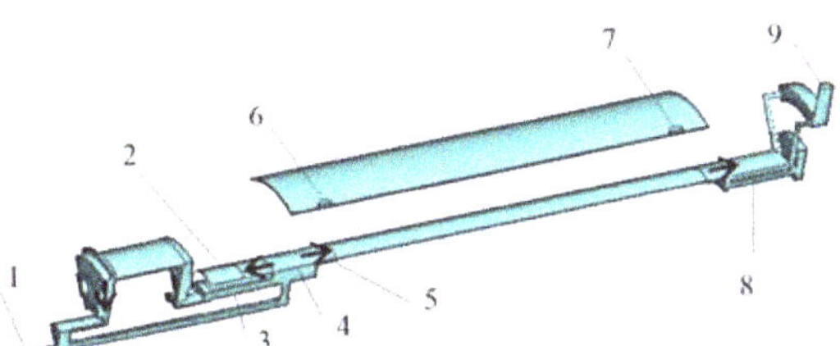

Figure 2. Schematic diagram of the water circulation circuit.

Considering the continuity of flow of water in the shield, the geometry characteristics of rotor and stator winding slots along the circumference, and the characteristics of heat transfer in stator and rotor material, the 1/8th structure of the motor along circumferential direction was simulated. The rotor shaft of the model was coincident with the z-coordinates; the positive direction of z-axis was towards down, and the y-axis was positive along radial direction. The coordinate origin point was located in the upper part and can be seen in Figure 1.

In Figure 2, 1 indicates the primary water Inlet, 2 the flywheel clearance inert zone, 3 the lower radial bearing clearance, 4 the auxiliary impeller position (confluence region), 5 the water in double cans clearance, 6 the secondary water exit, 7 the secondary water inlet, 8 the upper radial bearing clearance, and 9 the primary water outlet.

3. Solution Conditions

3.1. Basic Assumptions

1. The water in the motor was an incompressible fluid because the Mach number ($Ma = 0.24$) was far less than 0.7. The mass flow rate (5.127 kg/s) of primary water was very large in the clearance of the motor. The Reynolds's number ($Re = 5363$) showed that the flow was in the turbulence state.

2. Only the steady state of fluid flow was considered. There was no contact thermal resistance between the adjacent bodies. The flow of the fluid in the internal passage was in the smooth region because the internal surface roughness of the wall in the motor was very small and the friction loss can be uniformly implemented for the corresponding positions of the water as volumetric heat sources inside the fluid.

3. The electrical and harmonic losses and their distribution were calculated by another research institute using 3D software from the electromagnetic field. The corresponding loss in each body was uniformly distributed. Because the parameters of the thermo-physical properties of materials change with temperature variations, e.g., thermal conductivity, density, and specific heat at constant pressure, to be on the safe side, they were set as constant in the calculation process, so the minimum value of these parameters was chosen in the interval of the temperature through a large number of calculations. Among them, the thermal conductivity of the iron core lamination of rotor and stator was anisotropic, and the axial, radial, and tangential values of thermal conductivity were chosen experimentally. Except iron core lamination, other materials were isotropic, and their thermal conductivity values were selected conventionally.

3.2. Mathematical Model

The rotation of the water in the shield sleeve gap and the water in the center of the axis was considered in the calculation process because the rotor rotated rapidly relative to the stator under the condition of stable operation. Therefore, a multiple reference-frame system and the rotational speed of the wall surface were selected in the Ansys Fluent software (v14.5, Fluent Inc., New York, NY, USA, 2012). Three-dimensional flow and heat transfer coupling equations including the mass and momentum conservation equation and the relation of absolute velocity vector u (the velocity viewed from the stationary frame) and relative velocity vector u_r (the velocity viewed from the rotating frame) can be written as follows:

$$\nabla(\rho u_r) \tag{1}$$

$$\nabla(\rho u_r u_r) + \rho(2\Omega \times u_r + \Omega \times \Omega \times r) = -\nabla p + \nabla \tau + F) \tag{2}$$

$$u = u_r + \Omega \times r \tag{3}$$

In the fixed reference frame, mass, momentum, and energy conservation equations can be replaced by a general control equation, which is given in Equation (4).

$$\nabla(\rho u \varnothing) = \nabla(\Gamma\, grad\phi) + S \tag{4}$$

where Ω is the angular velocity of the rotor, ϕ is a universal variable, which can be replaced by unknown variables u, v, w, T, etc., Γ is the generalized diffusion coefficient, and S is the generalized source term. The only difference among the variables was the setting condition of Γ, S, initial values, and boundary conditions. The specific expressions of Γ, S for different variables were provided in [16]. In addition, for the solid components in the motor, the energy equation was converted into the heat conduction differential equation due to the no convection item. In the shielded motor, the mode of heat transfer was conduction between and inside the solid materials (e.g., stator core), and between the solid and liquid was the convection heat transfer. The natural convective heat transfer occurred between the nitrogen and the adjacent solid wall surfaces at the end of the wire rod in the upper and lower end closed room, and the magnitude of natural convection heat transfer was determined by the temperature or density difference. The flow of water in the shielded casing and outer casing was a mixed flow, and the natural convection in these regions can be neglected. Due to the small clearance between the stator and rotor shield sleeve and in the three bearings, the width of flow was in the order of millimeters. The forced convection flow in this region was affected by viscous shear stress and Coriolis force, an effect whereby a mass moving in a rotation system experiences a force acting perpendicular to the direction of motion and to the axis of rotation. Since a forced convection flow around a cylinder is dominated by viscous shear forces in the boundary layer and the velocity gradient between the walls is very large, so, in this paper, the shear stress transport (SST) k-ω two-equation mathematical model was adopted. This model included two transport equations to represent the turbulent properties of the flow. The first transported variable was turbulent kinetic energy k, and the second transported variable was ω. These transported variables determined the energy in turbulence and the scale of turbulence, respectively, and their equations can be written as follows:

$$\frac{\partial}{\partial x_i}(\rho k u_i) = \frac{\partial}{\partial x_j}\left(\Gamma_k \frac{\partial k}{\partial x_j}\right) + G_k - Y_k + S_k \tag{5}$$

$$\frac{\partial}{\partial x_i}(\rho \omega u_i) = \frac{\partial}{\partial x_j}\left(\Gamma_\omega \frac{\partial \omega}{\partial x_j}\right) + G_\omega - Y_\omega + S_\omega + D_\omega \tag{6}$$

In the above equations, G_k and G_ω represent the generation of k and ω due to mean velocity gradients. Γ_k and Γ_ω represent the effective diffusivity of k and ω, respectively. Y_k and Y_ω represent the dissipation of k and ω due to turbulence. D_ω represents the cross-diffusion term. S_k and S_ω are user-defined source terms [17].

3.3. Mesh Generation

Due to the complex geometry of the model, both structured and unstructured meshes were used, while the total number of cells was 14,099,447, and the value of equisized skew was less than 0.8 for the overall mesh. The mesh was refined into those regions where large velocity and pressure gradients were expected. In this paper, the grid independence solution was obtained after employing different meshing schemes, interval sizes, and calculations.

3.4. Loss and Boundary Conditions

3.4.1. Water Friction Loss

There were losses that caused the temperature rise in the motor, and in this paper, only water friction loss and relative influencing factors were given more attention. Principally, friction loss can be divided into loss along the path and local losses. The losses along the path were due to the frictional resistance caused by the fluid moving along the flow path and occurred in the fluid boundary layer where the velocity gradient was comparatively large. Local losses were due to fluid flow through various curved pipelines, for example, in elbows, valves, etc., due to the deformation of water flow, the change of direction, and speed redistribution, and the internal eddy current caused losses. The loss along the path was caused by viscous friction, and the frictional resistance coefficient was closely related to the surface roughness of the water channel, the state of flow, and the properties of the coolant.

In this paper, the value of surface roughness for each method was set the same. The analysis of friction losses in the water gap clearance can be estimated by the equation derived for rotating cylinders, which can be written as [18,19]:

$$P = k c_f \pi \rho \Omega^3 r^4 \tag{7}$$

where k is the roughness coefficient, where the value was 1.0 for smooth surface [4], ρ is the density of water, l is the length of the cylinder, and r is the rotor radius. The friction coefficient c_f depends on the radius of the cylinder, the radial air-gap length δ, and the Reynolds number. The calculated result of the water friction loss at every clearance position of the water channel can be seen in Table 1. In order to verify the accuracy of the calculation results, a friction loss experiment was also carried out. The experimental loss value was obtained by the loss-separation method [18] and was provided by the manufacturer.

Table 1. Four conditions of friction losses (kW) at different positions in the water clearance region.

Clearance Position	Assumed Value	Analysis Value	Test Value	Assumed Value
Shield sleeve	0	384	407	$407 \times 1.1\%$
Upper radial bearing	0	17	17	$17 \times 1.1\%$
Lower radial bearing	0	17	17	$17 \times 1.1\%$
Upper fly wheel	0	793	722	$722 \times 1.1\%$
Lower fly wheel	0	580	690	$690 \times 1.1\%$
Auxiliary wheel	0	7	7	$7 \times 1.1\%$
Total	0	1798	1860	2046

Note: The data in the Table 1 are the sum of local and along-the-way losses at each location.

Table 1 shows that the analysis value of the water friction loss in the water clearance (region) of the lower fly wheel and double cans was smaller than that of the test value by 110 kW and 23 kW, respectively, and of the upper flywheel was larger than the test value by 71 kW. Analysis of friction loss was smaller than the experimental friction loss by 62 kW, a relative difference of -3.33%; hence, the results were accurate (the total experimental friction loss was large). Based on previous experience and for the sake of operational safety, it was assumed that the actual operating condition (value) was higher than the test value by 10% in every term. The values of analysis and experimental water friction loss at each position were provided by the manufacturer; see Table 1. In order to separate the influence of water friction loss on temperature rise, a situation in which the water friction loss was assumed as zero was designed, as shown in the Table 1.

It was assumed that the friction losses at each location was evenly distributed. The value of water friction loss per unit volume as a source term (kW/m^3) was calculated according to Table 1, and these values were applied to the corresponding boundary volume of water, respectively. The CFD method was different from the above-mentioned conditions in Table 1. The friction loss in the clearance of water at different positions was calculated by the CFD FLUENT software. The specific way of selecting the viscous heating item was to select it in the "options" list, which is right under the turbulence model, and the wall roughness was set in the boundary conditions according to the different positions of the cans. All the related settings and parameters used in the CFD models above calculations are listed in Table 2. We can see the differences between the two calculation methods of water fiction loss.

Table 2. The settings and parameters in the computational fluid dynamics (CFD) models for temperature calculations. SST, shear stress transport.

	CFD	Assumed	Analysis	Test	Assumed Value
Models-viscous-(SST) k-ω	on	on	on	on	on
Models-energy	on	on	on	on	on
Options-viscous heating	on	off	off	off	off
Boundary condition-moving wall	set	-	-	-	-
Shear condition	no slip	-	-	-	-
Wall roughness (mm)	1.0	-	-	-	-
Source item in water clearance region (kW/m^3)	-	-	on	on	on

Note: "-" denotes no need to set any value.

3.4.2. Electrical Loss

The electric loss in the rotor was calculated in the iron core, the shield sleeve, the end of the coupling loop, and in the copper bar. Harmonic loss of the wire rod was also a part of this calculation. The electric loss in stator was calculated in the shield sleeve, the copper bar, and in the iron core. In particular, the stator and rotor core teeth and yoke loss were calculated separately, and the teeth were divided into 16 sections along the radial direction. In addition, the loss value of the conical ring, teeth, and yoke position of the stator finger plate was considered. The stray loss was calculated by the experimental formula developed by the factory for the same kind of shield motor. By using the same loss calculation method as used in [ ], the obtained calculated values of the main parts were as given in Table 3.

Table 3. Electrical loss values in the main solid components.

Components	Loss (kW)
Rotor copper bar	72.0
Rotor shield (casing)	262.57
Stator winding	44.54
Stator shield (casing)	920.58
Conical ring	28.53
Rotor copper bar	72.0
Rotor shield (casing)	262.57
Stator winding	44.54
Stator shield (casing)	920.58
Conical ring	28.53

It can be noted from the above table that the electrical loss was comparatively more in the rotor and stator shielding case. It was assumed that all electrical losses were distributed uniformly in the corresponding components.

3.4.3. Boundary Conditions

The inlet boundary condition of the volume flow rate of water was mainly determined by the intersection point of the pump and motor operating curves and partially determined by the geometry characteristics. The entrance of the primary water below the base and the entrance of the external cooling water in the upper part of the shielding casing (see Figure 2) were the velocity boundary conditions, and the magnitude of velocity was set experimentally. The rated rotating speed of the motor shaft was 1500 r/min.

The inlet-outlet temperature of the primary water was closely related to factors such as the magnitude of internal friction loss, the temperature of cooling water in the external heat exchanger (same as the secondary water temperature in the cooling jacket), and the velocity of water. Through the calculation of the heat balance and heat transfer of the internal motor and external heat exchanger, respectively, and between the internal motor and external heat exchanger, at the operating condition of the highest external cooling water temperature of 41 °C at entrance, we obtained the internal cooling water inlet temperatures, 60.38 °C, 61.65 °C, and 62.99 °C, at the three mentioned conditions. For the convenience of comparison, the temperature of the internal water inlet was set at the same temperature of 60.38 °C.

The gauge pressure at the outlet of the external water was set to 0 Pa. The mode of heat transfer in the outer surface of the shell was natural convection. For safety measures, the surface convective heat transfer coefficient was set to 1 W (m^2 K), and the ambient air temperature of the outer surface was 48.9 °C. The boundary condition at the top of the motor was constant heat flux, and its value was water friction loss per unit volume at the bottom fluid of the flywheel. The boundary conditions of the right and left surface of the calculation domain (see Figure 1) were linked periodically. In addition, the governing equations for the fluid and solid regions were solved by using FLUENT 14.5, which uses the finite volume method. A second order upwind scheme was employed for the discretization of all the variables, and the simple scheme was used for the pressure-velocity coupling. The standard wall function was used to deal with the viscosity lamination near the solid wall surface, and the dimensionless distance $y+$ met the requirements. After trying several different mesh topologies and sizes, the grid-independent solution was obtained.

4. Results and Discussion

The temperature field in the solution domain was obtained by means of numerical simulation of Equations (1)–(6) at the loss and boundary conditions mentioned above.

For the convenience of analysis, five straight lines (sampling lines) parallel to the rotation axis were drawn in order to get the axial variation law of physical quantity. The position of each line is shown in Figure 3, and the number of sampling lines required for the analysis is given in decreasing order of radii. The sampling surface was located at the radial cross-section of the stator slot; Sampling Lines 1 and 2 pass through the center of the stator lower and upper winding, respectively, and the start and end coordinates of the two lines are the same. Sampling Line 3 is located in the center of the clearance of the cooling water between the stator and rotor shield, and Sampling Lines 4 and 5 pass through the center of a rotor bar and the rotor core, respectively. This paper presents the analysis of the effect of water friction loss on the temperature of canned motor components such as the stator winding, rotor copper bar, rotor iron core, and on the internal water passage. The temperature analysis of different parts is described below.

Figure 3. Position of sampling lines and the surface in cross-section.

4.1. Temperature Analysis of the Stator Winding

The temperature distribution characteristics of the stator winding were the same in the axial direction, and the highest temperature point was located in the middle of the nose of the two ends of the stator winding (see Figure 4), resulting in a reduction of the heat by the conduction from the two ends of the stator to the middle, and was then delivered by the primary and secondary water convection. The position of peak temperature was not changed when calculated for four different conditions; assuming no fiction loss, analysis, experimental, and assuming 1.1-times the experimental value. The peak temperature increased gradually by 178.6 °C, 185.8 °C, 188.2 °C, and 192.0 °C, and the maximum temperature did not exceed the permitted temperature of 200 °C, i.e., no over-temperature.

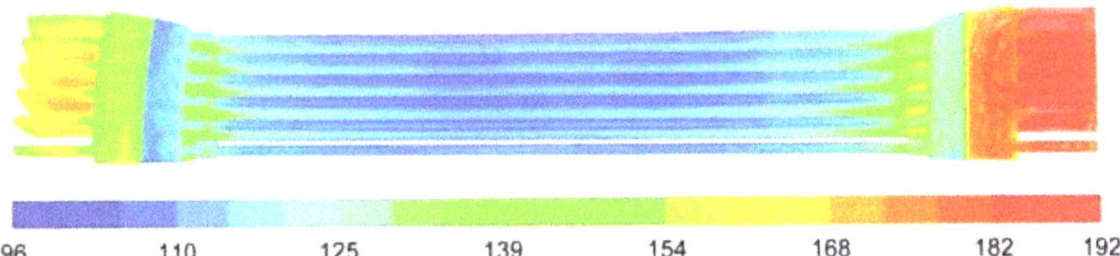

Figure 4. Temperature contour of the stator winding for the assumed maximum friction loss condition (°C).

The results showed that electrical loss was the main factor that rose the temperature of the winding, although the total value of the analysis of water friction loss was comparatively higher, but did not strongly affect the peak value of the temperature. Comparing it (peak temperature) with the no fiction loss condition, its temperature value increased by 7.2 °C, 9.6 °C, and 13.4 °C, and it showed that if the mechanical loss of water friction was neglected, the error would be about 10 °C. Comparing it (peak temperature) with the experimental calculation of the water friction loss condition, its temperature value increased by −2.4 °C and 3.8 °C, and the error was within the range ±2.0%, while the values of peak temperature increased non-linearly. The temperature contour of the stator winding at assumed 1.1-times the experimental calculation condition of water friction loss (maximum friction loss) is given in Figure 4.

In order to analyze the influence of water friction loss calculation methods on the temperature field of the motor, using the same conditions mentioned above and the CFD module, the obtained results showed that the peak temperature of the motor winding was 180.9 °C, and comparing with the experimental water friction loss values, it reduced by 7.93 °C (error was −3.84.2%), while temperature distribution characteristics and peak position did not change. The value of temperature was lower when calculated by the CFD module. The main cause of the error was that the actual wall roughness was higher than 1.0 mm.

It can be seen from Figure 4 that the peak temperature of the end center of the stator winding nose in the upper nitrogen end room was 192.0 °C. The highest temperature of stator winding in the lower nitrogen end room was lower than that in the upper part. The temperature of the winding adjacent to the central part of the iron core was lower because the surrounding water temperature in the internal gap and in external casing was lower and the convection heat transfer capability of the water was strong.

Figure 5 shows the temperature contour on the sampling surface for the test calculated condition. Theoretically, the temperature distribution characteristics on the surface of any polar angle along the circumferential cross-section are a typical case. The overall temperature distribution characteristics of any component along the radial direction can be seen in Figure 5. The temperature of the casing near the cooling water zones was low. Secondly, the temperature of the water in the stator and rotor shield sleeve clearance, stator iron core, and the rotor was higher; the stator winding temperature in the upper nitrogen chamber was the highest; and the temperature gradient near the tooth clamp plate was relatively large.

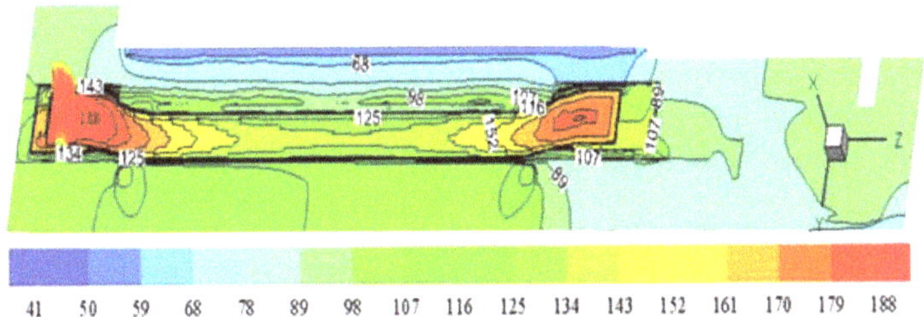

Figure 5. Temperature distribution of the radial sampling surface at a polar angle (°C).

Figures 6 and 7 show the temperature distribution of Sampling Lines 1 and 2 of the stator winding along the z-axis, respectively. From Figures 6 and 7, it can be seen that the temperature distribution of the winding presented a concave curve, and the lowest temperature point in the upper and lower windings was near the axial center of the windings. Ignoring frictional losses, the temperature rise caused by electric loss was shown in the curve corresponding to "0" in the figure, and this curve can be used as a benchmark for comparison. In comparison, it can be found that the temperature rise of the winding due to frictional loss at the upper and lower ends and the intermediate position was larger. It is one of the main purposes of this paper to reveal the influence of friction loss on the calculation of peak temperature. Except for the cavity at both ends, for the theoretical and experimental friction losses, the temperature distribution curve was basically coincident.

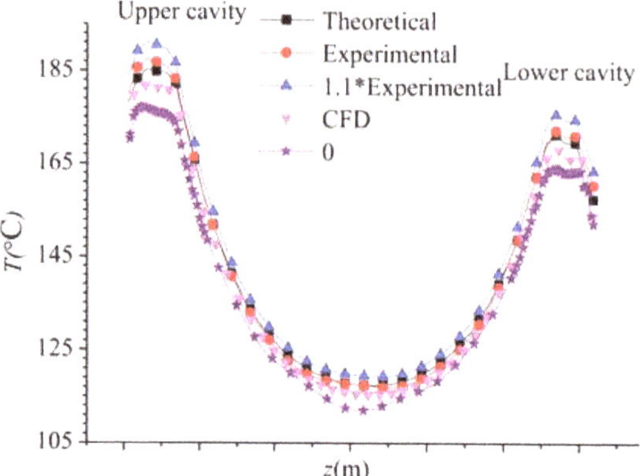

Figure 6. Temperature distribution of the lower part of stator winding Sampling Line 1.

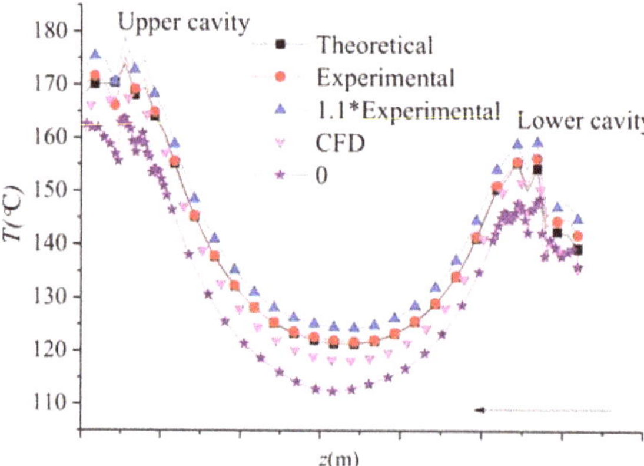

Figure 7. Temperature distribution of the upper part of stator winding Sampling Line 2.

The temperature distribution of the winding was not symmetrical along the axial center of the iron core. The temperature gradient in the left part of the winding was higher than that in the right part of the winding, and the winding temperature in the upper stator cavity was higher than that of the lower stator cavity. In the same z-coordinates, the temperature of Sampling Line 1 of the lower winding was higher than the temperature of Sampling Line 2. For Sampling Line 2, where the upper winding location was closer to the shaft, the actual length of the winding nose was short in the upper and lower part of the stator cavity.

In Figure 7, the irregular variation of the curve indicates the change of temperature in insulation, nitrogen (end room), and other parts through which the sampling line passed after stretching out the straight winding. Under conditions of analysis of and experimental water friction loss, the temperature of the winding was basically the same for the same position, except the end cavity. However, for the two end stator cavities, the calculation results were different in the five cases. The value of the winding temperature was the highest at the same position along the z-axis, under the water friction loss condition of 1.1-times the experimental value, and was lowest when calculated by the CFD module.

4.2. Temperature Analysis of the Primary Water Passage

Figure 8 shows the temperature distribution of Sampling Line 3 in the internal water passage between the stator and rotor shield. It can be seen from Figure 2 that Sampling Line 3, in turn, through the water clearance at the position of the upper guide bearing, double cans, near the confluence region, and approximately no water flowing area at the bottom of the lower flywheel, increased in the direction of the z-axis. The annular cross-sectional area of water in the confluence region became larger, whereas in the other region, the cross-sectional width was in mm.

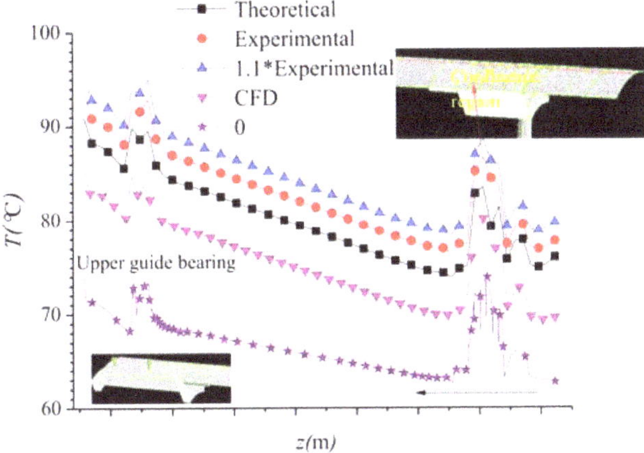

Figure 8. Temperature distribution of water passage Sampling Line 3 along the z-axis.

It can be seen in Figure 8 that the temperature of water rise was linear between the double cans region, and the cross-sectional area of water was the same. The maximum water temperature value for the four conditions of water friction loss was no more than the alarm temperature of 90 °C; however, the calculation results of the position of the upper guide bearing were inconsistent. The frictional loss did not exceed the temperature when calculated using CFD and the theoretical method, but in the other two cases, local temperature exceeded the rated condition. This example shows that the different calculation methods of friction loss had an influence on the conclusion about whether the water of the upper guide bearing was over temperature or not. It can provide a reference for the design of the internal water passage. The average temperature of water corresponding to the experimental friction loss value of Sampling Line 3 was higher by 2.6 °C than that of the analysis value calculated at the same position, and was lower by about 2.0 °C than the assumed value. For the CFD case, the average value of the temperature of Sampling Line 3 was lower by about 4.8 °C than that of the analysis value calculated at the same position. The average temperature of the water was 8.2 °C higher than that without friction loss.

The sudden rise of the temperature of two local positions is indicated in Figure 8. In the clearance (gap) of the upper guide bearing, the temperature of the water increased linearly, and the temperature of the water near the exit, at the bottom of the upper flywheel, increased dramatically due to the heat source in it. In the water clearance near the confluence region, there was an approximately no water flow region, so a lower rate of heat transfer was in this area. The water temperature was higher than that of the inlet of the rotor shield sleeve.

4.3. Temperature Analysis of the Rotor Copper Bar

Figure 9 shows the temperature variations of the sampling line of the rotor copper bar along the z-axis for the five conditions of water friction losses. It shows that the temperature distribution characteristics of the rotor cooper bar were different from the stator winding mentioned above in Figures 6 and 7. The temperature of the rotor copper strip at the lower and upper end loop position rose dramatically, and the temperature gradient was larger because there was a large heat source of electrical loss in the end ring.

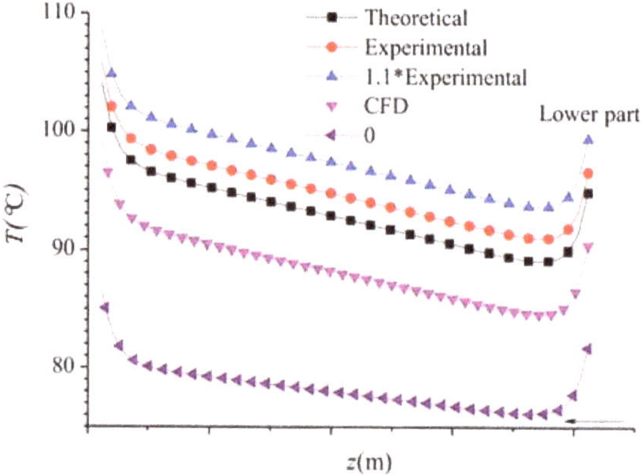

Figure 9. Temperature distribution of rotor copper bar Sampling Line 4.

The average temperature of the rotor copper bar of Sampling Line 4 corresponding to the analysis of friction loss value was lower by about 1.9 °C and 4.5 °C than that by the experimental value and by assuming 1.1-times the experimental value, respectively, and when friction losses were ignored, it was lower by 14.82 °C than that corresponding to the analysis of the friction loss value. The value of the average temperature of Sampling Line 4 was lower by about 4.71 °C than that of the calculated analysis value, when calculated by the CFD module. It shows that the large friction loss had a relatively large influence on the temperature rise of the rotor bar.

4.4. Temperature Analysis of the Rotor Iron Core

Figure 10 shows the temperature variations of Sampling Line 5 of the rotor iron core tooth along the z-axis for the five conditions of water friction losses, and it can be seen that the temperature distribution characteristics of rotor iron core were similar to Figure 9, different from Figures 6 and 7. There was a sharp rise in the local temperature gradient due to large electric loss in the tooth plate and in the end ring that were located in the lower and upper end of the rotor core, and the temperature distribution of the rotor iron core tooth changed linearly in the double cans section. The temperature of the rotor iron core also gradually increased, and the characteristics were the same as mentioned in the above Figure 9, while only the peak value was different.

It can be concluded that the characteristics of the temperature distribution curves did not change when the calculating method of water friction loss changed (as can be seen in Figures 6–10), and the value of the temperature (in Figures 8–10) was the highest at the same position under the water friction loss condition of 1.1-times the experimental value and lowest when calculated by the CFD module.

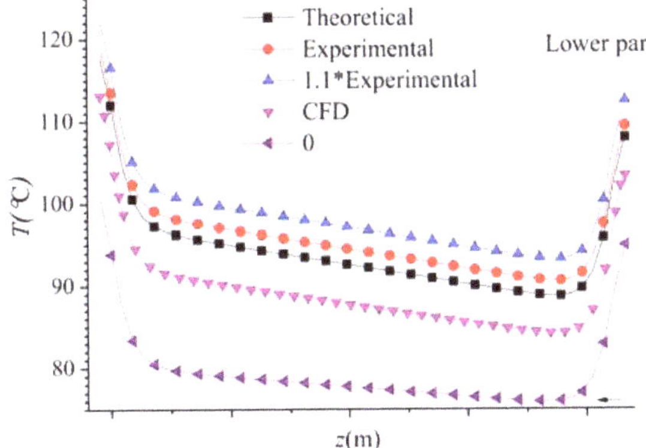

Figure 10. Temperature distribution (°C) of rotor core Sampling Line 5.

4.5. Accuracy Analysis of Temperature Calculation Results

The temperature calculations obtained at different water friction loss conditions had very small differences from each other. The position of peak temperature corresponding to the analysis value of the loss was same, but the peak value difference was −1.2 °C, when compared with the result of the earlier finite difference method software, which was developed by the factory.

Additionally, under the no load operating condition, multi-point temperature measurement had been performed. To verify the accuracy of the calculation results, the same numerical method (the water friction loss per unit volume was set as heat sources in the corresponding clearance of water) was adopted to calculate the temperature field under the no-load test condition. The error in the temperature results on the many positions simulated (using the method mentioned above) was 5% lower than the test values. However, at present, the thermal test of the motor in the fully-loaded condition has not yet been performed.

The results in this paper were accurate and satisfied the engineering requirements.

5. Conclusions

In this paper, a canned motor has been taken as an example, and a method was presented to study the effect of friction losses on the calculation of the thermal field of the motor using the CFD FLUENT software, that is: the water friction loss per unit volume was added in the corresponding position as the heat source; the viscous heating item option was kept inactive; and the wall roughness in the boundary conditions was not set. For the second method, the friction losses was calculated by the CFD FLUENT software, whereby the option of the viscous heating item was selected, which is right under the turbulence model list, and the wall roughness in the boundary conditions was set according to the different positions of the cans. The conclusion are as follows:

1. The peak temperature of the motor winding was lower by −4.2% as compared to the water friction loss calculated by CFD with the proposed water friction loss per unit volume method. This method of, i.e., using given experimental water friction loss per unit volume as the source term did not change the characteristics of the temperature distribution of the components of the motor and intercooling water.

2. The temperature along the axial direction increased with the increase of friction loss at the same position of the rotor copper bar, rotor iron core, and for water in the clearance of double cans, and the tendency of the temperature distribution increased linearly along the axial direction where the shield was located.

3. The temperature distribution of the winding presented a concave curve, and the lowest temperature point in the upper and lower winding was near the axial center of the winding.

4. For the water-cooled shielded motor, the water friction loss cannot be ignored. The calculation method of the water friction loss had a great influence on the conclusion about whether the water of the upper guide bearing was over temperature.

Author Contributions: Y.L. and A.M. are the main authors of this manuscript. All the authors contributed to this manuscript. All authors read and approved the final manuscript. Y.L. and A.M. conceived of the novel idea and performed the analysis; A.M. and Y.L. analyzed the data; A.M. and Y.L. contributed analysis tools; A.M. and Y.L. wrote the entire paper; M.A.R., S.R., S.A., M.S., and A.W.A. checked, reviewed, and revised the paper. Y.L. performed the final proofreading and supervised this research.

Funding: This research received no external funding.

Conflicts of Interest: The authors declare no conflict of interest.

References

1. Boglietti, A.; Cavagnino, A. Evolution and modern approaches for thermal analysis of electrical machines. *IEEE Trans. Ind. Electron.* **2009**, *56*, 871–882. [CrossRef]
2. Yetgin, A.G. Effects of induction motor end ring faults on motor performance. Experimental results. *Eng. Fail. Anal.* **2009**, *96*, 374–383. [CrossRef]
3. Liang, Y.; Bian, X.; Yu, H.; Li, C. Finite-element evaluation and eddy-current loss decrease in stator end metallic parts of a large double-canned induction motor. *IEEE Trans. Ind. Electron.* **2015**, *62*, 6779–6785. [CrossRef]
4. Huang, Z.; Fang, J.; Liu, X.; Han, B. Loss calculation and thermal analysis of rotors supported by active magnetic bearings for high-speed permanent-magnet electrical machines. *IEEE Trans. Ind. Electron.* **2016**, *63*, 2027–2035. [CrossRef]
5. Chiu, H.C.; Jang, J.H.; Yan, W.M.; Shiao, R.B. Thermal performance analysis of a 30 kW switched reluctance motor. *Int. J. Heat Mass Tran.* **2017**, *114*, 145–154. [CrossRef]
6. Kim, C.; Lee, K.S. Thermal nexus model for the thermal characteristic analysis of an open-type air-cooled induction motor. *Appl. Therm. Eng.* **2017**, *112*, 1108–1116. [CrossRef]
7. Kim, C.; Lee, K.S. Numerical investigation of the air-gap flow heating phenomena in large-capacity induction motors. *Int. J. Heat Mass Tran.* **2017**, *110*, 746–752. [CrossRef]
8. Lu, Y.; Liu, L.; Zhang, D. Simulation and analysis of thermal fields of rotor multislots for nonsalient-pole motor. *IEEE Trans. Ind. Electron.* **2015**, *62*, 7678–7686. [CrossRef]
9. Lancial, N.; Torriano, F.; Beaubert, F.; Harmand, S.; Rolland, G. Study of a Taylor-Couette-Poiseuille flow in an annular channel with a slotted rotor. In Proceedings of the International Conference on Electrical Machines (ICEM), Hangzhou, China, 2–5 September 2014; pp. 1422–1429.
10. Traxler-Samek, G.; Zickermann, R.; Schwery, A. Cooling airflow, losses, and temperatures in large air-cooled synchronous machines. *IEEE Trans. Ind. Electron.* **2010**, *57*, 172–180. [CrossRef]
11. Tu, J.; Yeoh, G.H.; Liu, C. *Computational Fluid Dynamics: A Practical Approach*, 3rd ed.; Butterworth-Heinemann: Oxford, UK, 2018.
12. Luomi, J.; Zwyssig, C.; Looser, A.; Kolar, J.W. Efficiency optimization of a 100-W 500,000-r/min permanent-magnet machine including air-friction losses. *IEEE Trans. Ind. Appl.* **2009**, *45*, 1368–1377. [CrossRef]
13. Gengji, W.; Ping, W. Rotor loss analysis of PMSM in flywheel energy storage system as uninterruptable power supply. *IEEE Trans. Appl. Superconduct.* **2016**, *26*, 1–5. [CrossRef]
14. Aglen, O.; Andersson, A. Thermal analysis of a high-speed generator. In Proceedings of the 38th IAS Annual Meeting on Conference Record of the Industry Applications Conference, Salt Lake City, UT, USA, 12–16 October 2003; Volume 1, pp. 547–554.
15. Le, Y.; Sun, J.; Han, B. Modeling and design of 3-DOF magnetic bearing for high-speed motor including eddy-current effects and leakage effects. *IEEE Trans. Ind. Electron.* **2016**, *63*, 3656–3665. [CrossRef]
16. Zhang, F.; Du, G.; Wang, T.; Wang, F.; Cao, W.; Kirtley, J.L. Electromagnetic design and loss calculations of a 1.12-MW high-speed permanent-magnet motor for compressor applications. *IEEE Trans. Energy Convers.* **2016**, *31*, 132–140. [CrossRef]
17. Lomax, H.; Pulliam, T.H.; Zingg, D.W. *Fundamentals of Computational Fluid Dynamics*, 1st ed.; Springer Science & Business Media: Berlin, Germany, 2013.

18. Heins, G.; Ionel, D.M.; Patterson, D.; Stretz, S.; Thiele, M. Combined experimental and numerical method for loss separation in permanent-magnet brushless machines. *IEEE Trans. Ind. Appl.* **2016**, *52*, 1405–1412. [CrossRef]
19. Saari, J. Thermal Analysis of High-Speed Induction Machines. Ph.D. Dissertation, Department of Electrical and Communications Engineering, Helsinki University of Technology, Espoo, Finland, 1998.